ISSN 0044-2267

ZEITSCHRIFT FÜR ANGEWANDTE MATHEMATIK UND MECHANIK

INGENIEURWISSENSCHAFTLICHE FORSCHUNGSARBEITEN

BAND 58 **1978** **HEFT 11**

AUS DEM INHALT

H A U P T A U F S Ä T Z E

A K A D E M I E - V E R L A G · B E R L I N

ZAMM **Bd. 58** **Nr. 11** **S. 477—528** **Berlin,** November **1978**

EVP 18,— M **34115**

INHALT

Wir bitten, **Manuskriptsendungen** zweifach (Original und eine Kopie, sprachlich einwandfrei, Formeln mit Maschine oder in Druckschrift geschrieben) an folgende Anschrift zu richten:

Zeitschrift für Angewandte Mathematik und Mechanik z. Hd. Herrn Prof. Dr. Günter Schmidt
Zentralinstitut für Mathematik und Mechanik an der Akademie der Wissenschaften der DDR
DDR-108 Berlin, Mohrenstraße 39.

Zu den Arbeiten, die als Hauptaufsätze bestimmt sind, ist auf gesondertem Blatt eine Zusammenfassung von 5 bis 10 Zeilen in englischer und (möglichst) deutscher und russischer Sprache beizufügen. Ausführliche Hinweise für die Autoren, um deren strikte Berücksichtigung gebeten wird, finden sich im Anschluß an das Inhaltsverzeichnis des Jahrganges **57** (1977).

Die Autoren erhalten von den Hauptaufsätzen 75, von den Kleinen Mitteilungen 25 Sonderdrucke ohne Berechnung, darüber hinaus weitere Sonderdrucke gegen Berechnung.

Der Verlag behält sich für alle Beiträge das Recht der Vervielfältigung, Verbreitung und Übersetzung vor.

Bezugsmöglichkeiten: Bestellungen sind zu richten

— **in der DDR** an den Postzeitungsvertrieb, an eine Buchhandlung oder an den Akademie-Verlag, DDR-108 Berlin, Leipziger Straße 3—4

— **im sozialistischen Ausland** an eine Buchhandlung für fremdsprachige Literatur oder an den zuständigen Postzeitungsvertrieb

in der BRD und Westberlin an eine Buchhandlung oder an die Auslieferungsstelle KUNST UND WISSEN, Erich Bieber, 7 Stuttgart 1, Wilhelmstraße 4—6

— **in Österreich** an den Globus-Buchvertrieb, 1201 Wien, Höchstädtplatz 3

— **im übrigen Ausland** an den Internationalen Buch- und Zeitschriftenhandel; den Buchexport, Volkseigener Außenhandelsbetrieb der Deutschen Demokratischen Republik, DDR-701 Leipzig, Postfach 160, oder an den Akademie-Verlag, DDR-108 Berlin, Leipziger Straße 3—4

ZEITSCHRIFT FÜR ANGEWANDTE MATHEMATIK UND MECHANIK
Herausgeber und Chefredakteur: Prof. Dr. Günter Schmidt (redaktionelle Mitarbeit: Dr. W. Heinrich und Dr. H. Weinert), Zentralinstitut für Mathematik und Mechanik der Akademie der Wissenschaften der DDR. Verlag: Akademie-Verlag, DDR-108 Berlin, Leipziger Straße 3—4; Fernruf: 2236229 oder 2236221; Telex-Nr.: 114420; Bank: Staatsbank der DDR, Berlin, Kto.-Nr.: 6836-26-20712. Anschrift der Redaktion: Zentralinstitut für Mathematik und Mechanik der Akademie der Wissenschaften, DDR-108 Berlin, Mohrenstraße 39; Fernruf: 2000561

Veröffentlicht unter der Lizenznummer 1282 des Presseamtes beim Vorsitzenden des Ministerrates der Deutschen Demokratischen Republik. Gesamtherstellung: VEB Druckerei „Thomas Müntzer", DDR-582 Bad Langensalza. Erscheinungsweise: Die Zeitschrift für Angewandte Mathematik und Mechanik erscheint monatlich. Die 12 Hefte eines Jahres einschließlich Tagungsheft bilden einen Band. Bezugspreis je Band 300,— M zuzüglich Versandspesen (Preis für die DDR 216,—M). Bezugspreis je Heft 25,— M (Preis für die DDR 18,—M). Bestellnummer dieses Heftes: 1009/58/11.

AN(EDV) 35937

ISSN 0044-2267

ZAMM 58, 477 – 481 (1978)

V. V. RUMYANTSEV / A. S. SUMBATOV

On the Problem of a Generalization of the Hamilton-Jacobi Method for Nonholonomic Systems

Es werden notwendige und hinreichende Bedingungen für die Anwendbarkeit des verallgemeinerten Hamilton-Jacobischen Integrationsverfahrens auf die Bewegungsgleichungen mechanischer Systeme mit nichtholonomen Zwangsbedingungen vom Chetaevschen Typ erhalten. Ein Beispiel wird betrachtet.

The necessary and sufficient conditions of the applicability of the generalized Hamilton-Jacobi method for integrating the equations of motion of mechanical systems with nonholonomic constraints of Chetaev's type are obtained. An example is considered.

Получены необходимые и достаточные условия применимости обобщенного метода Гамильтона-Якоби для интегрирования уравнений движения механических систем с неголономными связями типа Четаева. Рассмотрен пример.

1. Introduction

A well-known method of a complete integral has been developed for integrating canonical HAMILTON equations. In [1] C. G. JACOBI contemplated its modification for the equations of motion in the form with multipliers of finite constraints. Generalizing and extending this unfinished JACOBI's investigation G. K. SUSLOV in [2, 3] has given an integration method being based on the consideration of some undetermined multipliers and on the use of a complete integral of the generalized partial differential HAMILTON equation of the first order. I. S. ARZANYH in [4] and other authors investigated possibilities of applying this method to equations of motion with multipliers of linear nonholonomic constraints. It has been proved that for a nonholonomic system the method gives trajectories which in general contradict the law of motion. Moreover, some conditions have been obtained, under which, provided they are satisfied, any partial solution (sometimes, the general one) of the equations of motion of a nonholonomic system may be determined by means of the method (see the corresponding references in [5]).

Recently R. VAN DOOREN in a small paper [6] and then in an extensive report [7] has suggested a generalization of this method (without references to [1—3]) for nonholonomic systems with nonlinear constraints of CHETAEV's type and has applied it to some examples including the example of P. APPELL. The essence of the method consists in the following.

Let the *position of a mechanical system* be determined by the generalized coordinates q_j and let its *motion* be subject to nonholonomic ideal constraints of CHETAEV's type [8]

$$f_l(q_j, \dot{q}_j, t) = 0, \qquad (l = 1, \dots, r < n;\ j = 1, \dots, n) \tag{1.1}$$

such that the possible displacements δq_j of the system satisfy the relations

$$\frac{\partial f_l}{\partial \dot{q}_j}\,\delta q_j = 0, \qquad \left(\operatorname{rank}\left\|\frac{\partial f_l}{\partial \dot{q}_j}\right\| = r\right). \tag{1.2}$$

In addition it will be assumed that each function f_e is a homogeneous one of k_e-th order with respect to the generalized velocities $\dot{q}_j$. A dot supercript denotes differentiation with respect to time t. Summation over a repeated suffix is understood throughout this paper.

If $L(q_j, \dot{q}_j, t)$ is the Lagrangian of the system then the *generalized momenta* and the *Hamiltonian* of the system are defined by

$$p_j = \frac{\partial L}{\partial \dot{q}_j}, \qquad H(q_j, p_j, t) = p_j\dot{q}_j - L\,. \tag{1.3}$$

The *equations of motion* of the system may be presented in the canonical form

$$\dot{q}_j = \frac{\partial H}{\partial p_j}, \qquad \dot{p}_j = -\frac{\partial H}{\partial q_j} + \mu_e\frac{\partial f_e}{\partial \dot{q}_j}, \tag{1.4}$$

where μ_l are multipliers of the constraints (1.1). According to the generalized HAMILTON-JACOBI method the equations

$$\frac{\partial S}{\partial q_j} = p_j + \lambda_l\frac{\partial f_l}{\partial \dot{q}_j} \tag{1.5}$$

must be considered. Here λ_l are indefinite multipliers, S is a certain function. The equations (1.5) and the relations (1.4a) will allow us to express the multipliers λ_l as a function of $\frac{\partial S}{\partial q_j}$, p_j, q_j, t. Inserting these expressions for λ_l into (1.5) and using the constraint equations (1.1) we get a system of equations from which the momenta p_i will result as functions of $\frac{\partial S}{\partial q_j}$, q_j, t

$$p_j = p_j\left(\frac{\partial S}{\partial q_i}, q_i, t\right). \tag{1.6}$$

Substituting (1.6) instead of p_j in the Hamiltonian $H(q_j, p_j, t)$ of the system we write the modified partial differential equation of HAMILTON

$$\frac{\partial S}{\partial t} + H\left[q_j, p_j\left(\frac{\partial S}{\partial q_i}, q_i, t\right), t\right] = 0. \tag{1.7}$$

The following statement governs the generalized HAMILTON-JACOBI method [6, 7]: Let

$$S = S(q_j, t, \alpha_j) + c$$

be a complete integral of (1.7) *where* α_i *and* c *are arbitrary constants, then the motion of the system is obtained by solving the equations*

$$\frac{\partial S}{\partial \alpha_j} = \beta_j \qquad (\beta_j\text{-arbitrary constants}), \tag{1.8}$$

for q_i *and the momenta* p_i *are found from* (1.6).

In this paper we establish that in general this statement is not correct and expressions (1.6), (1.8) are the first integrals of the equations (1.4), (1.1) if and only if some conditions are satisfied.

2. Conditions of applicability of the generalized Hamilton-Jacobi method

It is more convenient for us to take the equations of motion in LAGRANGE's form

$$\frac{d}{dt}\frac{\partial T}{\partial \dot{q}_j} - \frac{\partial T}{\partial q_j} - \frac{\partial U}{\partial q_j} = \mu_e \frac{\partial f_e}{\partial \dot{q}_j}, \tag{2.1}$$

where $T = T_2 + T_1 + T_0$ is the kinetic energy and U is the force function of the system. Here T_2 denotes a quadratic form of the generalized velocities, T_1 is a linear form, T_0 and U don't depend on the velocities. From (1.1) and (2.1) the multipliers μ_e may be found as functions of $\dot{q}_i, q_i, t$. The equations (2.1) are equivalent to (1.4) in consequence of the convertibility of the transformation (1.3).

Let us introduce the new variables

$$\pi_j = \frac{\partial T}{\partial \dot{q}_j} + \lambda_l \frac{\partial f_l}{\partial \dot{q}_j}. \tag{2.2}$$

From these relations the generalized velocities $\dot{q}_j$ may be represented as functions of λ_l, π_i, q_i, t. Inserting these expressions of $\dot{q}_j$ into (1.1) we get a system of equations from which the multipliers λ_l will be obtained as functions of π_i, q_i, t. Substituting for λ_l in the expressions of q_j we get the equations

$$q_j = q_j(q_i, \pi_i, t). \tag{2.3}$$

It must be noted that the equations (2.3) are functionally dependet in consequence of (1.1); so the transformation (2.3) from π_i to $\dot{q}_j$ is not convertible.

Further one can notice that the function H in (1.7) may be obtained by substiting (2.3) for $\dot{q}_j$ in the Hamiltonian of the system

$$H = T_2 - T_0 - U = \tilde{H}(q_i, \pi_i, t) \tag{2.4}$$

and by the replacements (see (1.5) and (2.2))

$$\pi_j = \frac{\partial S}{\partial q_j}. \tag{2.5}$$

Now we'll show that *if* $S = S(q_j, t, \alpha_j) + c$ *is a complete integral of* (1.7) *and the variables* q_j, π_j *are defined by* (1.8), (2.5) *then the following* HAMILTON *system of equations*

$$q_j = \frac{\partial \tilde{H}}{\partial \pi_j}, \qquad \dot{\pi}_j = -\frac{\partial \tilde{H}}{\partial q_j} \tag{2.6}$$

holds (*the equations* (2.6a) *are another form of* (2.3)).

Really, differentiating (1.8) with respect to time we get

$$\frac{\partial^2 S}{\partial t\, \partial \alpha_j} + \frac{\partial^2 S}{\partial q_i\, \partial \alpha_j}\dot{q}_i = \frac{d\beta_j}{dt} = 0. \tag{2.7}$$

On the other hand, inserting $S = S(q_j, t, \alpha_j) + c$ into (1.7) an identity is obtained from which we get

$$\frac{\partial}{\partial \alpha_j}\left(\frac{\partial S}{\partial t} + \tilde{H}\right) = \frac{\partial^2 S}{\partial \alpha_j\, \partial t} + \frac{\partial \tilde{H}}{\partial \pi_i}\frac{\partial^2 S}{\partial \alpha_j\, \partial q_i} = 0. \tag{2.8}$$

From (2.7) and (2.8) we get the equations (2.6a) because of

$$\det\left\|\frac{\partial^2 S}{\partial \alpha_j\, \partial q_i}\right\| \neq 0.$$

The equations (2.6b) are obtained from (2.5) and (1.7) analogously:

$$\dot{\pi}_j = \frac{\partial^2 S}{\partial t\,\partial q_j} + \frac{\partial^2 S}{\partial q_i\,\partial q_j}\dot{q}_i = \frac{\partial^2 S}{\partial t\,\partial q_j} + \frac{\partial^2 S}{\partial q_i\,\partial q_j}\frac{\partial \tilde{H}}{\partial \pi_i} = -\frac{\partial \tilde{H}}{\partial q_j} + \frac{\partial}{\partial q_j}\left(\frac{\partial S}{\partial t} + \tilde{H}\right) = -\frac{\partial \tilde{H}}{\partial q_j}.$$

So one can assert that if the motion of the system may be obtained by solving (1.8) for q_i then the equations (2.6) are equivalent to the equations (2.1) (or to (1.4)) and vice versa. However, the following theorem is true.

Theorem 1: *A solution of equations* (2.6) *is also a solution of* (2.1) *if and only if it satisfies the relation*

$$\lambda_l\left(\frac{d}{dt}\frac{\partial f_l}{\partial \dot{q}_i} - \dot{q}_i\frac{\partial}{\partial q_j}\bigg|_\pi \frac{\partial f_l}{\partial \dot{q}_i}\right)\delta q_j = 0\,, \tag{2.9}$$

in which δq_j are possible displacements.

The sign $|_\pi$ points out that the partial derivative is calculated always supposing π_j are constants. The analogous convention will be met for the sign $|_{\dot{q}}$.

Proof: Calculating $d\pi_j/dt$ from (2.2) and (2.6b) we get

$$\frac{d}{dt}\frac{\partial T}{\partial \dot{q}_j} + \dot{\lambda}_l\frac{\partial f_l}{\partial \dot{q}_j} + \lambda_l\frac{d}{dt}\frac{\partial f_l}{\partial \dot{q}_j} = -\frac{\partial \tilde{H}}{\partial q_j}\bigg|_\pi. \tag{2.10}$$

But in accordance with (2.4) and (2.3) the right hand sides of these equations may be represented as

$$-\frac{\partial T_2}{\partial q_j}\bigg|_\pi + \frac{\partial}{\partial q_j}(T_0 + U) = -\frac{\partial T_2}{\partial q_j}\bigg|_{\dot{q}} - \frac{\partial T_2}{\partial \dot{q}_i}\frac{\partial \dot{q}_i}{\partial q_j}\bigg|_\pi + \frac{\partial}{\partial q_j}(T_0 + U)\,.$$

We have

$$\frac{\partial T_2}{\partial \dot{q}_i}\frac{\partial \dot{q}_i}{\partial q_j}\bigg|_\pi = \left(\pi_i - \frac{\partial T_1}{\partial \dot{q}_i} - \lambda_l\frac{\partial f_l}{\partial \dot{q}_i}\right)\frac{\partial \dot{q}_i}{\partial q_j}\bigg|_\pi = \frac{\partial}{\partial q_j}\bigg|_\pi(\pi_i\dot{q}_i) - \left(\frac{\partial T_1}{\partial \dot{q}_i} + \lambda_l\frac{\partial f_l}{\partial \dot{q}_i}\right)\frac{\partial \dot{q}_i}{\partial q_j}\bigg|_\pi$$
$$= \frac{\partial}{\partial q_j}\bigg|_\pi\left[\left(\frac{\partial T}{\partial \dot{q}_j} + \lambda_l\frac{\partial f_l}{\partial \dot{q}_i}\right)\dot{q}_i\right] - \left(\frac{\partial T_1}{\partial \dot{q}_i} + \lambda_l\frac{\partial f_l}{\partial \dot{q}_i}\right)\frac{\partial \dot{q}_i}{\partial q_j}\bigg|_\pi$$

in virtue of (2.2). Taking into account that $\frac{\partial f_l}{\partial \dot{q}_i}\dot{q}_i = k_l f_l$, $\frac{\partial T}{\partial \dot{q}_i}\dot{q}_i = 2T_2 + T_1$ (it is no summation over l) and the relations

$$\frac{\partial T_1}{\partial q_j}\bigg|_\pi = \frac{\partial T_1}{\partial q_j}\bigg|_{\dot{q}} + \frac{\partial T_1}{\partial \dot{q}_i}\frac{\partial \dot{q}_i}{\partial q_j}\bigg|_\pi$$

holds, we get

$$\frac{\partial T_2}{\partial \dot{q}_i}\frac{\partial \dot{q}_i}{\partial q_j}\bigg|_\pi = 2\frac{\partial T_2}{\partial q_j}\bigg|_\pi + \frac{\partial T_1}{\partial q_j}\bigg|_{\dot{q}} + \lambda_l\dot{q}_i\frac{\partial}{\partial q_j}\bigg|_\pi\frac{\partial f_e}{\partial \dot{q}_i}.$$

So the expressions (2.10) may be represented as follows

$$\frac{d}{dt}\frac{\partial T}{\partial \dot{q}_j} - \frac{\partial T}{\partial q_j}\bigg|_{\dot{q}} - \frac{\partial U}{\partial q_j} + \dot{\lambda}_l\frac{\partial f_l}{\partial \dot{q}_j} + \lambda_e\left(\frac{d}{dt}\frac{\partial f_l}{\partial \dot{q}_j} - \dot{q}_i\frac{\partial}{\partial q_j}\bigg|_\pi\frac{\partial f_l}{\partial \dot{q}_i}\right) = 0\,. \tag{2.11}$$

Necessity: If a solution of (2.6) also satisfies the equations (2.1) then the identities (2.11) yield

$$(\mu_e + \dot{\lambda}_e)\frac{\partial f_e}{\partial \dot{q}_i} + \lambda_e\left(\frac{d}{dt}\frac{\partial f_e}{\partial \dot{q}_j} - \dot{q}_i\frac{\partial}{\partial q_j}\bigg|_\pi\frac{\partial f_e}{\partial \dot{q}_i}\right) = 0 \tag{2.12}$$

from which the relation (2.9) follows on account of (1.2).

Sufficiency: Let a solution of (2.6) satisfy the relation (2.9). Then from (2.11) we get

$$\left(\frac{d}{dt}\frac{\partial T}{\partial \dot{q}_j} - \frac{\partial T}{\partial q_j} - \frac{\partial U}{\partial q_j}\right)\delta q_j = 0$$

and hence the equations (2.1) are obtained by virtue of (1.2). The theorem is proved.

Thus the generalized HAMILTON-JACOBI method of integration may be applied to the equations (1.4), (1.1) if and only if the relation (2.9) is true for the general solution of (2.6). Otherwise the functions q_j found from (1.8) don't form the general solution of (1.4), (1.1). But sometimes these functions may form a particular solution of (1.4), (1.1) if they satisfy (2.9), provided the constants α_i, β_i take particular values.

We can rewrite the relation (2.9) if the equations (1.1) of constraints are represented in the form

$$\dot{q}_a - \varphi_a(q_j, \dot{q}_b, t) = 0 \quad (a = 1, \dots, r;\ b, v = r + 1, \dots, n)$$

where the functions φ_a are homogeneous ones of first order with respect to $\dot{q}_{r+1}, \dots, \dot{q}_n$. Since $\delta q_a = \frac{\partial \varphi_a}{\partial \dot{q}_b}\delta q_b$ and δq_b are independent we get

$$\lambda_l\left[\dot{q}_v\left(\frac{\partial}{\partial q_a}\bigg|_\pi\frac{\partial \varphi_c}{\partial \dot{q}_v}\right)\frac{\partial \varphi_a}{\partial \dot{q}_b} - \left(\frac{d}{dt}\frac{\partial \varphi_l}{\partial \dot{q}_b} - \dot{q}_v\frac{\partial}{\partial q_b}\bigg|_\pi\frac{\partial \varphi_g}{\partial \dot{q}_v}\right)\right] = 0\,. \tag{2.13}$$

The first r relations of (2.12) which establish a connection between multipliers μ_e and λ_e become

$$\mu_a + \dot{\lambda}_a + \lambda_e \dot{q}_v \frac{\partial}{\partial q_a}\bigg|_\pi \frac{\partial \varphi_e}{\partial \dot{q}_v} = 0 . \tag{2.14}$$

The formulae (2.13) and (2.14) become more simplified when the kinetic energy T and all functions $\varphi_a(q_j, \dot{q}_b, t)$ don't depend on the variables $q_1, \dots, q_r$. Then we have

$$\lambda_l \left(\frac{\mathrm{d}}{\mathrm{d}t} \frac{\partial \varphi_l}{\partial q_b} - \dot{q}_v \frac{\partial}{\partial \dot{q}_b}\bigg|_\pi \frac{\partial \varphi_l}{\partial \dot{q}_v} \right) = 0 \tag{2.15}$$

and

$$\mu_a + \dot{\lambda}_a = 0 . \tag{2.16}$$

3. The case of linear constraints

Let the *motion of a mechanical system be subject to r ideal constraints*

$$f_l = A_{lj}(q_j, t)\, \dot{q}_j + A_l(q_i, t) = 0 \tag{3.1}$$

which are linear in the velocities (A_{lj}, A_l are the coefficients of these linear constraints). The *possible displacements* are definèd by $A_{lj}\delta q_l = 0$. Following [1—3] we consider the relations (1.5) and take the partial differential equation in the form

$$\frac{\partial S}{\partial t} - \lambda_e A_\varepsilon + \tilde{H}\left[q_j, p_j\left(\frac{\partial S}{\partial q_i}, q_i, t\right), t\right] = 0 \tag{3.2}$$

instead of (1.7). Then we have

$$\dot{q}_j = \frac{\partial \tilde{H}}{\partial \pi_j} - \frac{\partial \lambda_l}{\partial \pi_j} A_l , \qquad \dot{\pi}_j = -\frac{\partial \tilde{H}}{\partial q_j} + \frac{\partial}{\partial q_j}(\lambda_l A_l) \tag{3.3}$$

instead of (2.6). Looking at the proof of Theorem 1 we see that it must be somewhat modified for the considered case. Then the identities

$$\frac{\mathrm{d}}{\mathrm{d}t}\frac{\partial T}{\partial \dot{q}_j} - \frac{\partial T}{\partial q_j}\bigg|_{\dot{q}} - \frac{\partial U}{\partial q_j} + \dot{\lambda}_l A_{lj} + \lambda_l \left(\frac{\mathrm{d}A_{lj}}{\mathrm{d}t} - \dot{q}_i \frac{\partial A_{li}}{\partial q_j} - \frac{\partial A_l}{\partial q_j} \right) = 0 \tag{3.4}$$

are obtained instead of (2.11) and the following theorem can be proved.

Theorem 2: *Let $S = S(q_j, t, \alpha_j) + c$ be a complete integral of* (3.2). *The functions q_j found from* (1.8) *form the general law of motion of the mechanical system which is subject to the constraints* (3.1) *if and only if the general solution of* (3.3) *satisfies the relation*

$$\lambda_l \left[\left(\frac{\partial A_{lj}}{\partial q_i} - \frac{\partial A_{li}}{\partial q_j} \right) \dot{q}_i + \frac{\partial A_{lj}}{\partial t} - \frac{\partial A_l}{\partial q_j} \right] \delta q_j = 0 . \tag{3.5}$$

In the case of finite constraints $f_e(q_j, t)$ we have

$$A_{lj} = \frac{\partial f_l}{\partial q_j}, \qquad A_l = \frac{\partial f_l}{\partial t}$$

and the relation (3.5) *is satisfied identically. This fact is in agreement with* SUSLOV's *theory* [2, 3]. *Moreover, from* (3.4) *we get*

$$\dot{\lambda}_l = -\mu_l .$$

SUSLOV has called λ_e "*impulsive multipliers of constraints*".

If the equations of the linear constraints are given in the form

$$\dot{q}_m = a_{mv} q_v + a_m, \qquad (m = 1, \dots, r;\ b, v = r + 1, \dots, n) ,$$

and hence

$$\delta q_m = a_{mv} \delta q_v$$

then from (3.5) we get

$$\lambda_l \left[\left(\frac{\partial a_{lb}}{\partial q_m} a_{mv} + \frac{\partial a_{lb}}{\partial q_v} - \frac{\partial a_{lv}}{\partial q_m} a_{mb} - \frac{\partial a_{lv}}{\partial q_b} \right) \dot{q}_b + \frac{\partial a_{lv}}{\partial t} - \frac{\partial a_l}{\partial q_v} \right] = 0 . \tag{3.6}$$

In addition the first r relations of (3.4) yield

$$\mu_m + \dot{\lambda}_m + \lambda_l \left(\frac{\partial a_{lv}}{\partial q_m} \dot{q}_v - \frac{\partial a_l}{\partial q_m} \right) = 0 . \tag{3.7}$$

For the case of *scleronomic* mechanical systems the conditions (3.6) and the relations (3.7) have been obtained in [5] in a different way.

4. Example

In APPELL's example the equation of the nonlinear constraints may be represented as

$$\dot{q}_1 = a\sqrt{\dot{q}_2^2 + \dot{q}_3^2} = \varphi(\dot{q}_2, \dot{q}_3) \qquad (a = \text{const})\,, \tag{4.1}$$

and the LAGRANGE function of the system is

$$L = \frac{m}{2}(\dot{q}_1^2 + \dot{q}_2^2 + \dot{q}_3^2) - mgq_1\,.$$

The equations (2.2) for this example yield

$$\pi_1 = m\dot{q}_1 + \lambda,\ \pi_2 = m\dot{q}_2 - \lambda a\frac{\dot{q}_2}{\sqrt{\dot{q}_2^2 + \dot{q}_3^2}}, \qquad \pi_3 = m\dot{q}_3 - \lambda a\frac{\dot{q}_3}{\sqrt{\dot{q}_2^2 + \dot{q}_3^2}}$$

from which we determine

$$\dot{q}_1 = \frac{1}{m}(\pi_1 - \lambda),\ \dot{q}_2 = \frac{1}{m}\cdot\frac{\pi_2(\pi_1 - \lambda)}{\pi_1 - \lambda(1 + a^2)}, \qquad \dot{q}_3 = \frac{1}{m}\cdot\frac{\pi_3(\pi_1 - \lambda)}{\pi_1 - \lambda(1 + a^2)}.$$

Substituting $\dot{q}_j$ by these expressions in (4.1) we find

$$\lambda = \frac{\pi_1 - a\sqrt{\pi_2^2 + \pi_3^2}}{1 + a^2}$$

and hence

$$\dot{q}_1 = \frac{a}{m(1 + a^2)}\left(a\pi_1 + \sqrt{\pi_2^2 + \pi_3^2}\right), \qquad \dot{q}_2 = \frac{\pi_2\left(a\pi_1 + \sqrt{\pi_2^2 + \pi_3^2}\right)}{m(1 + a^2)\sqrt{\pi_2^2 + \pi_3^2}},$$
$$\dot{q}_3 = \frac{\pi_3\left(a\pi_1 + \sqrt{\pi_2^2 + \pi_3^2}\right)}{m(1 + a^2)\sqrt{\pi_2^2 + \pi_3^2}}\,. \tag{4.2}$$

The HAMILTON function in $q - \pi$ variables is

$$\tilde{H} = \frac{1}{2m(1 + a^2)}\left(a\pi_1 + \sqrt{\pi_2^2 + \pi_3^2}\right)^2 + mgq_1\,.$$

For the considered example the equations (2.6a) are (4.2) and the equations (2.6b) are

$$\dot{\pi}_1 = -mg, \qquad \dot{\pi}_2 = 0, \qquad \dot{\pi}_3 = 0\,. \tag{4.3}$$

The case $\lambda = 0$ is of no interest because it means the reaction force equals zero on account of (2.16). So we can represent the conditions (2.15) in the following from

$$\frac{\mathrm{d}}{\mathrm{d}t}\frac{\partial\varphi}{\partial\dot{q}_v} - \dot{q}_b\frac{\partial}{\partial q_v}\bigg|_\pi\frac{\partial\varphi}{\partial\dot{q}_b} = 0\,, \qquad (b, v = 2, 3)\,. \tag{4.4}$$

The general solution of (4.3) satisfies (4.4) since

$$\frac{\partial\varphi}{\partial\dot{q}_2} = \frac{\pi_2}{\sqrt{\pi_2^2 + \pi_3^2}}, \qquad \frac{\partial\varphi}{\partial\dot{q}_3} = \frac{\pi_3}{\sqrt{\pi_2^2 + \pi_3^2}}.$$

Therefore according to theorem 1 the general law of motion of the system may be determined by means of the generalized HAMILTON-JACOBI method. This fact has been established in [6, 7] directly.

It must be noted that the conditions (4.4) may not be satisfied in the case of other forces. Let for example the force function be q_2. Then we get the equations (4.2) and the equations

$$\dot{\pi}_1 = 0, \qquad \dot{\pi}_2 = 1, \qquad \dot{\pi}_3 = 0 \tag{4.5}$$

instead of (4.3). It is evident that the conditions (4.4) are not satisfied for the general solution of (4.5). But they are satisfied for the particular solution of (4.5) in which $\pi_3 = 0$, i.e. $q_3 = \text{const}$.

References

1 JACOBI, C. G., Vorlesungen über Dynamik von JACOBI nebst fünf hinterlassenen Abhandlungen desselben, herausgegeben von A. CLEBSCH. Nachgelassene Abhandlungen, Berlin, 1866, S. 376–380.

2 Суслов, Г. К., Об уравнениях с частными производными для несвободного движения, Санкт-Пемербург, 1888.

3 Суслов, Г. К., Теоретическая механика, М.-Л., Гостехиздат, 1944, стр. 461–477.

4 Аржаных, И. С., Условия применимости потенциального метода интегрирования уравнений движения неголономных консервативных систем, Докл. АН СССР **87**, 15–18 (1952).

5 Сумбатов, А., С., Об интегрировании уравнений динамики с множителями связей, Прикл. матем. и механика **36**, 163–171 (1972).

6 VAN DOOREN, R., The generalized HAMILTON-JACOBI method for nonholonomic dynamical systems of CHETAEV's type, ZAMM **55**, 407–411 (1975).

7 VAN DOOREN, R., Generalized methods for nonholonomic systems with applications in various fields of classical mechanics, Theor. and Appl. Mech. 14th IUTAM Congr., Delft. 1976.

8 Четаев, Н. Г., О принципе Гаусса, Изв. Казанского физико-матем. общества 6, 68–71 (1933).

Eingereicht: 10. 6. 1977

Anschriften: V. V. RUMYANTSEV, Corresponding Member of Academy of Sciences of the USSR; A. S. SUMBATOV, Junior Research Worker of Computer Centre of Academy of Sciences of the USSR, Computer Centre of the Academy of Sciences of the USSR, Vavilov Street 40, Moscow, USSR

ZAMM 58, 483 – 487 (1978)

K. Gröger

Zur Theorie des dynamischen Verhaltens von elastisch-plastischen Körpern

In dieser Arbeit wird gezeigt, daß das dynamische Verhalten von elastisch-plastischen Körpern unter gewissen Annahmen durch Anfangswertprobleme für Evolutionsungleichungen zweiter Ordnung beschrieben werden kann. Für ein spezielles Materialgesetz wird die eindeutige Lösbarkeit derartiger Aufgaben bewiesen.

In this paper it is shown that under certain assumptions the dynamic behaviour of elastic-plastic materials can be described by initial value problems for second order differential inequalities. In the case of a special constitutive relation the unique solvability of such problems is proved.

В статье показано, что динамическое поведение эласто-пластических тел при некоторых предположениях можно описать с помощью задач Коши для эволюционных уравнений второго порядка. Для одного частного материального уравнения доказано однозначная разрешимость таких задач.

0. Einführung

Die vorliegende Arbeit ist eine Fortsetzung der Arbeit [4], in der wir das quasi-statische Verhalten von elastisch-plastischen Körpern behandelt haben. Im ersten Abschnitt formulieren wir mathematische Aufgabenstellungen, die das dynamische Verhalten eines dreidimensionalen Körpers beschreiben. Unser Ausgangspunkt ist dabei eine Modellvorstellung, wie sie von Del Piero [2] angegeben worden ist. Die sich ergebenden Aufgaben sind Rand-Anfangswertproblemen für partielle Differentialgleichungen sehr ähnlich. Im zweiten Abschnitt beweisen wir für ein spezielles Materialgesetz (das auf Prager zurückgeht) die Korrektheit der genannten Aufgabenstellungen, d. h. die Existenz genau einer Lösung und deren stetige Abhängigkeit von Anfangswerten und äußeren Kräften.

Wir werden in dieser Arbeit ohne nochmalige Erläuterung von den Begriffen und Bezeichnungen Gebrauch machen, die wir in den Abschnitten 1 und 2 der Arbeit [4] eingeführt haben. Insbesondere werden wir den dort erklärten Begriff des *mechanischen Prozesses* benutzen, der in gewissem Sinne das Verhalten eines elastisch-plastischen Körpers charakterisiert. Hinweise auf die Formeln (1), (2), ... aus [4] werden wir in der Form (I.1), (I.2), ... schreiben.

1. Formulierung von Aufgabenstellungen

Unser Ziel ist es, in diesem Abschnitt Aufgabenstellungen herzuleiten, die einen dynamischen Verformungsprozeß eines dreidimensionalen elastisch-plastischen Körpers bei gegebenen äußeren Kräften beschreiben. Wie in [4] nehmen wir der Einfachheit halber an, daß der Körper eingespannt ist. Es ist ohne weiteres möglich, auch andere Randbedingungen zu betrachten. Wir gehen davon aus, daß das Materialgesetz des betrachteten Körpers durch die Beziehungen (I.8), (I.9) charakterisiert wird.

Im folgenden sei u_0 eine Anfangsverschiebung, v_0 eine Anfangsgeschwindigkeit, σ_0 eine mit u_0 physikalisch verträgliche Anfangsspannung und f ein Kraftprozeß, d. h., es gelte

$$u_0 \in U, \qquad v_0 \in U, \qquad \sigma_0 \in C(Ku_0) = C_0 + BKu_0, \qquad f \in H^1(S; U^*) .$$

Die Bestimmung des dynamischen Verhaltens des elastisch-plastischen Körpers führt bei Vorgabe der genannten Größen auf die Suche nach einem *Verschiebungsprozeß* $u \in H^1(S; U)$ und einem *Spannungsprozeß* $\sigma \in H^1(S; \Sigma)$ mit folgenden Eigenschaften:

$$u'' + K^*\sigma = f , \tag{1}$$

$$[Ku, \sigma] \text{ ist ein mechanischer Prozeß} , \tag{2}$$

$$[u(0), u'(0), \sigma(0)] = [u_0, v_0, \sigma_0]; \tag{3}$$

dabei bezeichnet u'' die zweite Ableitung von u im Distributionensinne. Ist (1) mit $\sigma \in L^2(S; \Sigma)$ und $f \in L^2(S; U^*)$ erfüllt, so gehört u'' automatisch zum Raum $L^2(S; U^*)$. Die Gleichung (1) beinhaltet die Gleichgewichtsbeziehungen für den Körper (vgl. (I.19)). Wir haben der Einfachheit halber angenommen, daß der Körper die konstante Dichte 1 besitzt. Die Aussage (2) besagt, daß die für das Verhalten des Körpers charakteristische Beziehung zwischen Deformationen und Spannungen erfüllt ist.

In [4] haben wir gesehen, daß (2) gleichwertig ist zu

$$\sigma = (\Lambda_{\tau_0} + \partial I_{C_0})^{-1} \big((A - B)\, Ku\big)' + BKu, \quad \tau_0 = \sigma_0 - BKu_0 \tag{4}$$

(vgl. (I.17)). Auf Grund der Definition von Λ_{τ_0} läßt sich (4) auch in der folgenden Form schreiben:

$$\sigma' + \partial I_{C_0}(\sigma - BKu) \ni (AKu)', \qquad \sigma(0) = \sigma_0 . \tag{5}$$

Die Aufgabe (1)—(3) ist daher äquivalent zu

$$u'' + K^*\sigma = f , \tag{6}$$

$$\sigma' + \partial I_{C_0}(\sigma - BKu) \ni (AKu)' , \tag{7}$$

$$[u(0), u'(0), \sigma(0)] = [u_0, v_0, \sigma_0] , \tag{8}$$

$$u \in H^1(S; U), \qquad \sigma \in H^1(S; \Sigma) . \tag{9}$$

Setzt man den in (4) angegebenen Wert für σ in die Gleichung (1) ein, so erhält man eine Aufgabenstellung, in der explizit nur die Verschiebung u auftritt:

$$\begin{aligned} &u'' + K^*\big((A_{\tau_0} + \partial I_{C_0})^{-1}((A - B)Ku)' + BKu\big) = f\,, \\ &u(0) = u_0, \quad u'(0) = v_0, \quad \tau_0 = \sigma_0 - BKu_0, \quad u \in H^1(S; U)\,. \end{aligned} \tag{10}$$

Im Fall *quasi-statischer Verformungen* hatte es sich als zweckmäßig herausgestellt, eine Formulierung der Aufgabe in den Spannungen anzugeben. Diese Formulierung war der mathematischen Behandlung relativ leicht zugänglich. Wir werden im folgenden zeigen, daß im Falle *dynamischer Verformungen* eine Formulierung der Aufgabe günstig ist, die sich aus (6)—(9) ergibt, wenn man anstelle von σ die neue unbekannte Funktion

$$t \to \zeta(t) := \int_0^t (\sigma - BKu)(s)\, ds \quad \text{für alle} \quad t \in s$$

einführt. Unter Benutzung von ζ läßt sich die Aufgabe (6)—(9) wie folgt schreiben:

$$u'' + K^*(\zeta' + BKu) = f\,, \tag{11}$$

$$\zeta'' + \partial I_{C_0}\zeta' \ni ((A - B)Ku)'\,, \tag{12}$$

$$[u(0), u'(0), \zeta(0), \zeta'(0)] = [u_0, v_0, 0, \sigma_0 - BKu_0]\,, \tag{13}$$

$$u \in H^1(S; U), \qquad \zeta \in H^1(S; \Sigma)\,. \tag{14}$$

Hierbei verstehen wir ∂I_{C_0} als mehrdeutige Abbildung von $L^2(S; \Sigma)$ in sich. Aus (12) folgt daher insbesondere, daß ζ'' dem Raum $L^2(S; \Sigma)$ angehört.

Bemerkung 1: Die vorangegangenen Überlegungen zeigen, daß die Aufgabenstellungen (1)—(3) und (11) bis (14) äquivalent sind. Kennt man eine Lösung $[u, \zeta]$ der Aufgabe (11)—(14), so kann man die Spannungen σ aus der Beziehung $\sigma = \zeta' + BKu$ ermitteln.

2. Lösung der Aufgaben im Spezialfall

Wie im Fall quasi-statischer Verformungen sind wir leider nicht in der Lage, die im vorigen Abschnitt erhaltenen Aufgabenstellungen unter den dort genannten Voraussetzungen zu lösen. Ähnlich wie in [4], Abschnitt 3, spezialisieren wir deshalb das Materialgesetz, indem wir annehmen, daß die in (I.8), (I.9) eingeführten Abbildungen A und B folgendermaßen gegeben sind:

$$A = J, \qquad B = kJ, \qquad 0 < k < 1; \tag{15}$$

dabei bezeichnet J den durch (I.2) definierten Isomorphismus vom Raum $\mathscr{E}$ der Deformationen auf den Raum $\mathscr{S}$ der Spannungen. Die Aufgabe (6)—(9) erhält unter der Voraussetzung (15) die Gestalt

$$u'' + K^*\sigma = f\,, \tag{16}$$

$$\sigma' + \partial I_{C_0}(\sigma - kJKu) \ni JKu'\,, \tag{17}$$

$$[u(0), u'(0), \sigma(0)] = [u_0, v_0, \sigma_0], \tag{18}$$

$$u \in H^1(S; U), \qquad \sigma \in H^1(S; \Sigma)\,. \tag{19}$$

Analog ergibt sich anstelle von (10) bzw. (11)—(14):

$$\begin{aligned} &u'' + K^*\big((A_{\tau_0} + \partial I_{C_0})^{-1}(1 - k)JKu' + kJKu\big) = f\,, \\ &u(0) = u_0, u'(0) = v_0, \quad \tau_0 = \sigma_0 - kJKu_0, \quad u \in H^1(S; U) \end{aligned} \tag{20}$$

bzw.

$$u'' + K^*(\zeta' + kJKu) = f\,, \tag{21}$$

$$\zeta'' + \partial I_{C_0}\zeta' \ni (1 - k)JKu' \tag{22}$$

$$[u(0), u'(0), \zeta(0), \zeta'(0)] = [u_0, v_0, 0, \sigma_0 - kJKu_0]\,, \tag{23}$$

$$u \in H^1(S; U), \qquad \zeta \in H^1(S; \Sigma)\,. \tag{24}$$

Um auf das Problem (21)—(24) einen bekannten Satz über Evolutionsgleichungen zweiter Ordnung anwenden zu können, führen wir die folgenden HILBERT-Räume ein:

$$H := L^2(G; R^3), \qquad U_1 := U \times \Sigma, \qquad H_1 := H \times \Sigma, \qquad U_1^* := U^* \times \Sigma\,.$$

Für $g \in U^*$ und $u \in U$ (bzw. $g \in H$, $u \in H$) bezeichnen wir mit (g, u) den Wert des linearen Funktionals g im Punkt u (bzw. das übliche Skalarprodukt von g und u im Raum H). Es erweist sich als zweckmäßig, für $[g, \tau] \in U_1^*$ und $[u, \zeta] \in U_1$ (bzw. $[g, \tau] \in H_1$, $[u, \zeta] \in H_1$) das folgende Skalarprodukt einzuführen:

$$([g, \tau], [u, \zeta])_1 := (1 - k)(g, u) + (\tau, \zeta)_\Sigma\,.$$

U_1^* kann im Sinne des eben eingeführten Skalarprodukts als der zu U_1 duale Raum angesehen werden. Dabei wird automatisch H_1 mit seinem dualen Raum identifiziert.

Wir definieren nun $L \in \mathscr{L}(U_1, U_1^*)$ durch die Beziehung

$$L[u, \zeta] := k[K^*JKu, 0] \quad \text{für alle} \quad [u, \zeta] \in U_1\,. \tag{25}$$

Unter Benutzung von (I.18) ergibt sich

$$(L[u,\zeta],\ [u,\zeta])_1 + \|[u,\zeta]\|_{H_1}^2 \geqq (1-k)\ k(K^*JKu,\ u) + \|\zeta\|_\Sigma^2$$
$$= (1-k)\ k\|Ku\|_\Gamma^2 + \|\zeta\|_\Sigma^2 \geqq (1-k)\ kk_0\|u\|_U^2 + \|\zeta\|_\Sigma^2 \geqq c(\|u\|_U^2 + \|\zeta\|_\Sigma^2) = c\|[u,\zeta]\|_{U_1}^2, \quad c > 0\,. \tag{26}$$

Neben L benötigen wir eine mehrdeutige Abbildung M von U_1 in U_1^*, die wie folgt definiert werden kann:

$$M[u,\zeta] = \{[K^*\zeta,\tau]\,|\tau + (1-k)\ JKu \in \partial I_{C_0}\zeta\} \quad \text{für alle} \quad [u,\zeta] \in U_1\,. \tag{27}$$

Mit $D(M)$ bezeichnen wir, wie üblich, die Menge derjenigen Paare $[u,\zeta] \in U_1$, für die $M[u,\zeta] \neq \emptyset$ ist. Offensichtlich ist $D(M) = U \times C_0$.

Lemma 1: *Die durch* (27) *definierte Abbildung M ist maximal monoton.*

Beweis: Es sei $u_i \in U$, $\zeta_i \in C_0$ und $[K^*\zeta_i, \tau_i] \in M[u_i,\zeta_i]$ für $i = 1, 2$. Dann gilt auf Grund der Definition von M und der Monotonie von ∂I_{C_0} als mehrdeutige Abbildung von Σ in sich

$$([K^*(\zeta_1-\zeta_2),\ \tau_1-\tau_2],\ [u_1-u_2,\ \zeta_1-\zeta_2])_1 = (1-k)\ (K^*(\zeta_1-\zeta_2),\ u_1-u_2) + (\tau_1-\tau_2,\ \zeta_1-\zeta_2)_\Sigma$$
$$= \big(\tau_1-\tau_2+(1-k)\ (JKu_1-JKu_2),\ \zeta_1-\zeta_2\big)_\Sigma \geqq 0\,.$$

Diese Ungleichung beweist die Monotonie von M. Es gelte nun

$$([g-K^*\zeta,\ \sigma-\tau],\ [u_1-u,\ \zeta_1-\zeta])_1 \geqq 0 \quad \text{für alle} \quad [u,\zeta] \in D(M) \quad \text{und alle} \quad [K^*\zeta,\tau] \in M[u,\zeta]\,.$$

Die Maximalität von M ist bewiesen, wenn wir zeigen können, daß hieraus die Beziehung $[g,\sigma] \in M[u_1,\zeta_1]$ folgt. Die letzte Ungleichung ist äquivalent zu

$$(1-k)\ (g-K^*\zeta,\ u_1-u) + \big(\sigma-\tilde\tau+(1-k)\ JKu,\zeta_1-\zeta\big)_\Sigma \geqq 0 \text{ für alle } \tilde\tau \in \partial I_{C_0}\zeta, \text{ alle } u \in U \tag{28}$$
$$\text{und alle } \zeta \in C_0\,.$$

Wählt man $u = u_1$, so erhält man

$$\big(\sigma+(1-k)\ JKu_1-\tilde\tau,\ \zeta_1-\zeta\big)_\Sigma \geqq 0 \quad \text{für alle} \quad \tilde\tau \in \partial I_{C_0}\zeta \quad \text{und alle} \quad \zeta \in C_0\,.$$

Auf Grund der maximalen Monotonie von ∂I_{C_0} folgt daraus $\zeta_1 \in C_0$ und

$$\sigma + (1-k)\ JKu_1 \in \partial I_{C_0}\zeta_1\,. \tag{29}$$

Folglich darf in (28) auch $\zeta = \zeta_1$ gewählt werden. Damit ergibt sich

$$(g-K^*\zeta_1,\ u_1-u) \geqq 0 \quad \text{für alle} \quad u \in U\,.$$

Das ist nur möglich, wenn $g = K^*\zeta_1$ ist. Aus dieser Beziehung und (29) folgt $[g,\sigma] \in M[u_1,\zeta_1]$. Damit ist die Maximalität des Operators M bewiesen.

Mit Hilfe der Operatoren L und M läßt sich die Aufgabe (21)—(24) folgendermaßen schreiben:

$$\begin{aligned} &[u'',\zeta''] + M[u',\zeta'] + L[u,\zeta] \ni [f,0]\,,\\ &[u'(0),\ \zeta'(0)] = [v_0,\ \sigma_0-kJKu_0],\ [u(0),\zeta(0)] = [u_0,0]\,,\\ &[u,\zeta] \in H^1(S;\ U_1)\,. \end{aligned} \tag{30}$$

Satz 1: *Es gelte*

$$u_0 \in U,\ v_0 \in U,\ \sigma_0 \in C_0 + kJKu_0, \quad K^*\sigma_0 \in H, \quad f \in H^1(S;H)\,. \tag{31}$$

Dann besitzt die Aufgabe (30) *genau eine Lösung.*

Beweis: Die Behauptung von Satz 1 folgt aus einem von Barbu angegebenen Satz über Evolutionsgleichungen zweiter Ordnung (siehe [1], Th. 1.1, Ch. V). Um den Satz anwenden zu können, hat man nachzuprüfen, daß folgendes gilt:

$$[f,0] \in H^1(S;H_1),\ [v_0,\ \sigma_0-kJKu_0] \in D(M)\,,$$
$$[kK^*JKu_0,0] + [K^*(\sigma_0-kJKu_0),\ \tau+(1-k)\ JKu_0] \in H_1 \quad \text{für ein } \tau \in \partial I_{C_0}(\sigma_0-kJKu_0)\,.$$

Da $D(M) = U \times C_0$ und $0 \in \partial I_{C_0}(\sigma_0-kJKu_0)$ ist, folgen diese Relationen aus der Voraussetzung (31). Für die Anwendung des Satzes benötigt man außerdem die Eigenschaft (26) des Operators L und die in Lemma 1 nachgewiesene maximale Monotonie von M. Damit ist Satz 1 bewiesen.

Folgerung: *Unter der Voraussetzung* (31) *sind die Aufgaben* (16)—(19) *und* (20) *eindeutig lösbar.*

Bemerkung 2: Satz 1 zeigt, daß für das durch (15) charakterisierte spezielle Materialgesetz folgendes gilt: Gibt man Anfangsverschiebung, Anfangsgeschwindigkeit, Anfangsspannung und äußere Kräfte vor, so existiert genau ein dynamischer Verformungsprozeß, wenn

a) Anfangsverschiebung und Anfangsspannung physikalisch verträglich sind (d. h. $\sigma_0 \in C_0 + kJKu_0$),
b) die vorgegebenen Größen gewissen Regularitätsvoraussetzungen genügen ($v_0 \in U$, $K^*\sigma_0 \in H$, $f \in H^1(S;H)$).

Wir werden später sehen, daß man sich von den unter b) aufgeführten Regularitätsforderungen — die nicht in der Natur der Aufgabenstellung liegen — weitgehend frei machen kann, indem man einen schwächeren, aber physikalisch durchaus sinnvollen Lösungsbegriff für die Aufgabe (16)—(19) einführt.

Im folgenden Satz befassen wir uns mit der Korrektheit der behandelten Aufgabenstellungen.

Satz 2: *Für $i = 1, 2$ gelte $f_i \in H^1(S; U^*) + L^2(S; H)$ und*

$$u_i'' + K^*\sigma_i = f_i\,, \tag{32}$$

$$\sigma_i' + \partial I_{C_0}(\sigma_i - kJKu_i) \ni JKu_i'\,, \tag{33}$$

$$u_i \in H^1(S; U), \qquad \sigma_i \in H^1(S; \Sigma)\,. \tag{34}$$

Dann ist

$$\|u_1'' - u_2''\|_{L^2(S;U^*)} + \|u_1' - u_2'\|_{C(S;H)} + \|u_1 - u_2\|_{C(S;U)} + \|\sigma_1 - \sigma_2\|_{C(S;\Sigma)}$$
$$\leqq \text{const}\left(\|(u_1' - u_2')(0)\|_H + \|(u_1 - u_2)(0)\|_U + \|(\sigma_1 - \sigma_2)(0)\|_\Sigma + \|f_1 - f_2\|_{H^1(S;U^*)+L^2(S;H)}\right).$$

(Zur Definition des Raums $H^1(S; U^*) + L^2(S; H)$ siehe etwa [3], Kap. I, § 5).

Beweis: Wir setzen zur Abkürzung $\tau_i := \sigma_i - kJKu_i$, $i = 1, 2$, und $u := u_1 - u_2$, $\sigma := \sigma_1 - \sigma_2$, $f := f_1 - f_2$, $\tau := \tau_1 - \tau_2$. Es sei $f = g + h$ mit $g \in H^1(S; U^*)$, $h \in L^2(S; H)$. Aus (33) folgt

$$\tau_i' + (k-1)\, JKu_i' + \partial I_{C_0}\tau_i \ni 0\,.$$

Daher gilt unter den Voraussetzungen (32)—(34) für jedes $t \in S$

$$0 \geqq \int_0^t \{(1-k)\big((u'' + K^*(\tau + kJKu) - g - h)(s), u'(s)\big) + \big((\tau' + (k-1)\,JKu')(s), \tau(s)\big)_\Sigma\}\, ds$$
$$= (1-k)\,\{\tfrac{1}{2}(\|u'(t)\|_H^2 - \|u'(0)\|_H^2 + k\|Ku(t)\|_\Gamma^2 - k\|Ku(0)\|_\Gamma^2) +$$
$$+ \int_0^t \big(g'(s), u(s)\big)\, ds - \big(g(t), u(t)\big) + \big(g(0), u(0)\big) - \int_0^t \big(h(s), u'(s)\big)\, ds\} +$$
$$+ \tfrac{1}{2}\|\tau(t)\|_\Sigma^2 - \tfrac{1}{2}\|\tau(0)\|_\Sigma^2\,.$$

Unter Benutzung von (I.18) folgt daraus

$$\|u'(t)\|_H^2 + \|u(t)\|_U^2 + \|\tau(t)\|_\Sigma^2 \leqq$$
$$\text{const}\left(\|u'(0)\|_H^2 + \|u(0)\|_U^2 + \|\tau(0)\|_\Sigma^2 + \|g\|_{C(S;U^*)}^2 + \int_0^t \{\|g'(s)\|_{U^*}^2 + \|u(s)\|_U^2 + \|h(s)\|_H^2 + \|u'(s)\|_H^2\}\, ds\right).$$

Mit Hilfe des GRONWALLschen Lemmas und der Beziehung $\tau = \sigma - JKu$ erhält man daraus weiter mit einer Konstanten c, die nicht von t und nicht von der gewählten Darstellung von f als Summe $g + h$, $g \in H^1(S; U^*)$, $h \in L^2(S; H)$, abhängt:

$$\|u'(t)\|_H^2 + \|u(t)\|_U^2 + \|\sigma(t)\|_\Sigma^2 \leqq c(\|u'(0)\|_H^2 + \|u(0)\|_U^2 + \|\sigma(0)\|_\Sigma^2 + \|g\|_{H^1(S;U^*)}^2 + \|h\|_{L^2(S;H)}^2)\,.$$

Da

$$\|f\|_{H^1(S;U^*)+L^2(S;H)} = \inf_{\substack{g+h=f\\ g \in H^1(S;U^*),\, f \in L^2(S;H)}} (\|g\|_{H^1(S;U^*)} + \|h\|_{L^2(S;H)})$$

ist, folgt daraus die Behauptung des Satzes, abgesehen von der Abschätzung für $\|u''\|_{L^2(S;U^*)}$. Wegen

$$u'' = f - K^*\sigma$$

ergibt sich die gewünschte Abschätzung für $\|u''\|_{L^2(S;U^*)}$ sofort aus dem vorangegangenen Resultat. Damit ist Satz 2 bewiesen.

Wir wollen nun, wie angekündigt, für die Aufgabenstellung (16)—(19) einen schwächeren Lösungsbegriff einführen. Es sei

$$D := \{[u_0, v_0, \sigma_0, f] \in U \times U \times \Sigma \times H^1(S; H) \mid \sigma_0 \in C_0 + kJKu_0,\ K^*\sigma_0 \in H\}\,.$$

Man überzeugt sich leicht davon, daß D in

$$X := \{[u_0, v_0, \sigma_0, f] \in U \times H \times \Sigma \times \big(H^1(S; U^*) + L^2(S; H)\big) \mid \sigma_0 \in C_0 + kJKu_0\}$$

dicht liegt, wenn man X mit der Metrik des Raums $U \times H \times \Sigma \times \big(H^1(S; U^*) + L^2(S; H)\big)$ versieht. Weiter sei

$$Y := \{[u, \sigma] \in C(S; U) \times C(S; \Sigma) \mid u' \in C(S; H),\ u'' \in L^2(S; U^*)\}$$

und

$$\|[u, \sigma]\|_Y := \|u\|_{C(S;U)} + \|u'\|_{C(S;H)} + \|u''\|_{L^2(S;U^*)} + \|\sigma\|_{C(S;\Sigma)}\,.$$

Der Inhalt der Sätze 1 und 2 läßt sich nun wie folgt zusammenfassen:

Zu jedem $[u_0, v_0, \sigma_0, f] \in D$ existiert genau eine Lösung $[u, \sigma]$ der Aufgabe (16)—(19), *und die Zuordnung $[u_0, v_0, \sigma_0, f] \mapsto [u, \sigma]$ ist als Abbildung aus X in Y LIPSCHITZ-stetig.*

Da D in X dicht liegt, kann die Zuordnung $[u_0, v_0, \sigma_0, f] \mapsto [u, \sigma]$ in eindeutiger Weise zu einer auf dem ganzen Raum X definierten LIPSCHITZ-stetigen Abbildung forgesetzt werden. Wir wollen diese Abbildung mit R bezeichnen.

Definition: Für beliebiges $[u_0, v_0, \sigma_0, f] \in X$ bezeichnen wir das Paar $[u, \sigma] = R[u_0, v_0, \sigma_0, f] \in Y$ als *schwache Lösung* des Problems (16)—(19).

Nach Definition von R ist $[u, \sigma] = R[u_0, v_0, \sigma_0, f]$ genau dann, wenn eine Folge $([u_{0i}, v_{0i}, \sigma_{0i}, f_i])$, $i = 1, 2, \ldots$, aus D existiert, für die gilt: $[u_{0i}, v_{0i}, \sigma_{0i}, f_i] \to [u_0, v_0, \sigma_0, f]$ in X, $[u_i, \sigma_i] \to [u, \sigma]$ in Y. Dabei bezeichnet $[u_i, \sigma_i]$ die nach Satz 1 eindeutig bestimmte Lösung der Aufgabe

$$u_i'' + K^*\sigma_i = f_i\,, \tag{35}$$

$$\sigma_i' + \partial I_{C_0}(\sigma_i - kJKu_i) \ni JKu_i'\,, \tag{36}$$

$$[u_i(0), u_i'(0), \sigma_i(0)] = [u_{0i}, v_{0i}, \sigma_{0i}]\,, \tag{37}$$

$$u_i \in H^1(S; U), \qquad \sigma_i \in H^1(S; \Sigma)\,. \tag{38}$$

Aus der Beziehung (35) folgt durch Grenzübergang in $L^2(S; U^*)$

$$u'' + K^*\sigma = f .$$

Das Paar $[u, \sigma]$ genügt also in der üblichen Weise der Gleichgewichtsbedingung. Die Funktionen u und σ erfüllen außerdem die Anfangsbedingungen $u(0) = u_0$, $u'(0) = v_0$, $\sigma(0) = \sigma_0$. Für eine schwache Lösung $[u, \sigma]$ sind die vom Standpunkt der Mechanik interessanten Größen wie Verschiebungen, Verschiebungsgeschwindigkeiten, Deformationen und Spannungen wohldefiniert. Die Frage nach dem dynamischen Verhalten eines Körpers (mit dem hier betrachteten Materialgesetz) kann also auch für Daten $[u_0, v_0, \sigma_0, f] \in X$ als sinnvoll angesehen werden.

In der Beziehung (36) kann man im allgemeinen nicht zur Grenze übergehen, da u' nicht zu $L^2(S; U)$ gehören muß. Man kann daher nicht zeigen, daß $[Ku, \sigma]$ ein mechanischer Prozeß im Sinne der Definition 2 in [4] ist. Diese Feststellung legt die Frage nahe, ob es möglich ist, den Begriff „mechanischer Prozeß" so zu erweitern, daß auch für jede schwache Lösung der Aufgabe (16)—(19) das Paar $[Ku, \sigma]$ automatisch ein mechanischer Prozeß ist. Leider ist uns eine derartige Erweiterung bisher nicht gelungen.

Literatur

1 Barbu, V., Nonlinear Semigroups and Differential Equations in Banach Spaces, Bukarest, Leyden 1976.
2 Del Piero, G., On the elastic-plastic material element, Arch. Rat. Mech. Analysis 59, 111—129 (1975).
3 Gajewski, H.; Gröger, K.; Zacharias, K., Nichtlineare Operatorgleichungen und Operatordifferentialgleichungen, Berlin 1974.
4 Gröger, K., Zur Theorie des quasi-statischen Verhaltens von elastisch-plastischen Körpern, ZAMM 58, 81—88 (1978).

Eingereicht: 2. 12. 1976

Anschrift: Dr. K. Gröger, AdW der DDR, Zentralinstitut für Mathematik und Mechanik, Mohrenstraße 39, DDR 108 Berlin

ZAMM 58, 489—499 (1978)

H. Bergander

Eine verallgemeinerte Darstellung inelastischer Deformationsgesetze zur Erleichterung der numerischen Lösung von Anfangs-Randwertproblemen

Es wird gezeigt, daß der Zusammenhang zwischen Spannung und Verzerrung für die unterschiedlichsten Materialklassen in eine einheitliche Form gebracht werden kann. Für die Zuwüchse von Spannung und Verzerrung entspricht diese Form der eines linearen Elastizitätsgesetzes mit Zusatzverzerrungen. Die Koeffizienten des Gesetzes hängen vom Spannungs- und Verzerrungszustand und internen Parametern ab. Die Zahl der Parameter ist für unterschiedliches Materialverhalten unterschiedlich groß. Die Formulierung erweist sich als günstig für eine numerische Lösung von Anfangs-Randwertproblemen, deren allgemeine Aspekte herausgearbeitet werden.

Показывается, что связь между напряжениями и деформациями для самых различных класс материала может быть приведена в единую форму. Для приращений напряжений и деформаций эта форма соответствует закону линейной упругости с дополнительными деформациями. Коэффициенты этого закона зависят от состояния напряжений и деформаций и от внутренних параметров. Число параметров для различного поведения материала отличается друг от друга. Формулировка оказывается удачной для численного решения начальных краевых задач, общие аспекты которых разрабатываются.

It is demonstrated that the connection between stress and strain can be presented in a united form for material classes of most various kinds. This form corresponds to a form of a linear elasticity law with additional strains for the increments of stress and strain. The coefficients of law depend on stress strain state and internal parameters. The number of parameters is differently great for various material states. This formulation is very advantageous for a numerical solution of boundary-value problems with initial conditions of which are worked out the general aspects.

1. Einleitung

Mit den zunehmenden Anforderungen an die Funktionsfähigkeit, die Sicherheit zum Schutze von Menschenleben bei hochbelasteten Aggregaten (z. B. der Kerntechnik) und an den minimalen Materialeinsatz zur Gewährleistung der jeweiligen Funktion wird der Zwang zur genaueren Berücksichtigung des Materialverhaltens bei der Lösung von Festigkeitsproblemen immer stärker. Zugleich hat die Entwicklung der elektronischen Rechentechnik bedeutend dazu beigetragen, daß heute Probleme mit numerischen Methoden umfassend gelöst werden können, die noch vor wenigen Jahrzehnten als unlösbar galten.

Auf dem Gebiet der Festkörpermechanik führt die genauere Analyse des Materialverhaltens meist in den Bereich der inelastischen Erscheinungen. Diese sind charakterisiert durch die Abhängigkeit des Spannungs-Verformungs-Zusammenhanges von der Vorgeschichte und von der Zeit sowie durch die Nichtlinearität dieses Zusammenhanges. Der Zusammenhang zwischen Spannung und Verzerrung im Materialelement soll als Deformationsgesetz bezeichnet werden. Der häufig hierfür angewendete Begriff Stoffgesetz wird vermieden, da auch die Beschreibung der zulässigen Beanspruchungen sowie eine Reihe weiterer Gesetzmäßigkeiten zur Charakterisierung des physikalischen Verhaltens spezieller Materialien Stoffgesetze darstellen.

Oft nimmt der Typ des Deformationsgesetzes eine Schlüsselstellung bei der Wahl des Lösungsverfahrens für die Aufgabe der Bestimmung des Spannungs-Verformungsfeldes unter der Wirkung äußerer Lasten ein. Die Vielfalt der Formen derartiger Deformationsgesetze läßt auch die Vielfalt der Lösungsmethoden zumindest in formaler Hinsicht stark anwachsen. Dabei fällt es dem Ingenieur, der ein spezielles Problem lösen muß, oft nicht leicht, zu erkennen, daß sich hinter diesem Formenreichtum nur wenige prinzipiell unterschiedliche Lösungsmethoden verbergen.

Anliegen dieser Arbeit soll es sein, eine formale Schreibweise von Deformationsgesetzen aufzuzeigen, in die nahezu alle bedeutsamen Deformationsgesetze der Festkörpermechanik entweder sofort oder nach einer technischen Anforderungen in der Regel voll genügenden Approximation einzuordnen sind. Diese Form ist ein Satz von Differentialoperatoren erster Ordnung bezüglich der Zeit, in dem eine bestimmte Anzahl von Zustandsparametern enthalten ist. Damit wird die Aufgabe der Ermittlung der Spannungs-Verformungsfelder auf ein Anfangs-Randwertproblem zurückgeführt. Dessen Besonderheit besteht darin, daß die Zuwüchse aller Zustandsgrößen bei gegebenem Zustand einem linearen Randwertproblem genügen. Das ermöglicht dann zugleich, ein einheitliches numerisches Lösungsverfahren für alle Typen des physikalischen Verhaltens auf der Grundlage von Methoden, die vor allem in der Plastizitätstheorie entwickelt worden sind, anzugeben.

2. Das allgemeine Deformationsgesetz

Mit kleinen stark ausgezogenen lateinischen und griechischen Buchstaben sollen Spaltenmatrizen (Vektoren), mit entsprechenden großen Buchstaben Matrizen gekennzeichnet werden. Ein sehr allgemeines Deformationsgesetz für inelastische Stoffe läßt sich dann in folgender Form angeben:

$$\boldsymbol{\varepsilon} = \boldsymbol{J}(\boldsymbol{\varepsilon}, \boldsymbol{\sigma}, \boldsymbol{h})\,\dot{\boldsymbol{\sigma}} + \boldsymbol{b}(\boldsymbol{\varepsilon}, \boldsymbol{\sigma}, \boldsymbol{h}), \qquad \dot{\boldsymbol{h}} = \boldsymbol{A}(\boldsymbol{\varepsilon}, \boldsymbol{\sigma}, \boldsymbol{h})\,\boldsymbol{\sigma} + \boldsymbol{c}(\boldsymbol{\varepsilon}, \boldsymbol{\sigma}, \boldsymbol{h})\,. \tag{1), (2}$$

Hierin ist $\boldsymbol{\varepsilon}$ der Vektor aller im jeweiligen konkreten Problem vorkommenden Komponenten des Verzerrungstensors, $\boldsymbol{\sigma}$ analog der Vektor der Komponenten des Spannungstensors, $\boldsymbol{h}$ ist ein Satz von Zustandsparametern. Die Matrizen $\boldsymbol{J}$ und $\boldsymbol{A}$ sowie die Vektoren $\boldsymbol{b}$ und $\boldsymbol{c}$ hängen vom jeweiligen Zustand, charakterisiert durch $\boldsymbol{\varepsilon}$, $\boldsymbol{\sigma}$ und $\boldsymbol{h}$ ab. Durch ihre spezielle Form wird der jeweilige Werkstoff charakterisiert. Der Punkt gibt die Ableitung nach der Zeit t an. Wenn die Werkstoffunktionen $\boldsymbol{J}$, $\boldsymbol{A}$, $\boldsymbol{b}$ und $\boldsymbol{c}$ zusätzlich explizit von t abhängen, wird das Einwirken nichtmechanischer Prozesse (z. B. chemische Vorgänge) auf das Material erfaßt. Die Zusammenfassung solcher Einflüsse wird als Alterung bezeichnet. Außerdem werden alle Werkstoffunktionen im allgemeinen sehr wesentlich von der Temperatur beeinflußt. Im nichtisothermen Fall ist dann z. B. (1) durch

$$\boldsymbol{\varepsilon} = \boldsymbol{J}(\boldsymbol{\varepsilon}, \boldsymbol{\sigma}, \boldsymbol{h}, T)\,\dot{\boldsymbol{\sigma}} + \boldsymbol{b}(\boldsymbol{\varepsilon}, \boldsymbol{\sigma}, \boldsymbol{h}_T, T) + \boldsymbol{b}(\boldsymbol{\varepsilon}, \boldsymbol{\sigma}, \boldsymbol{h}, T)\,T \tag{3}$$

zu ersetzen[1]), wobei sowohl die Änderung der Materialeigenschaften als auch die thermische Dehnung erfaßt werden muß. Im Rahmen dieser Arbeit soll ausschließlich alterungsfreies Material im isothermen Fall Berücksichtigung finden. Diese Festlegung vereinfacht die Schreibarbeit, während die Erweiterung keine Grundsatzprobleme aufwirft und leicht nachzuholen ist.

Für ein klassisches Kontinuum sind die Vektoren $\boldsymbol{\varepsilon}$ und $\boldsymbol{\sigma}$ maximal sechsdimensional. Im reduzierten Fall (z. B. ebener Spannungszustand, ebener Hauptspannungszustand usw.) können $\boldsymbol{\sigma}$ und $\boldsymbol{\varepsilon}$ prinzipiell unterschiedliche Dimension haben. So erzeugt z. B. ein ebener Spannungszustand $\boldsymbol{\sigma} = \{\sigma_{11}, \sigma_{22}, \sigma_{12}\}^T$ in isotropem Material $\boldsymbol{\varepsilon} = \{\varepsilon_{11}, \varepsilon_{22}, \varepsilon_{33}, \varepsilon_{12}\}^T$. Daher ist die Matrix $\boldsymbol{J}$ nicht immer quadratisch. Der Aufbau der Relation (1) für ein konkretes Berechnungsproblem kann jedoch stets so gewählt werden, daß nur korrespondierende Elemente auftreten. Das soll im weiteren zur Vereinfachung der Formeln vorausgesetzt werden.

Bezüglich $\dot{\boldsymbol{\varepsilon}}$ und $\dot{\boldsymbol{\sigma}}$ ist (1) dann sehr einfach invertierbar:

$$\boldsymbol{\sigma} = \boldsymbol{C}\boldsymbol{\varepsilon} + \boldsymbol{d} \tag{4}$$

mit

$$\boldsymbol{C} = \boldsymbol{J}^{-1}, \qquad \boldsymbol{d} = -\boldsymbol{J}^{-1}\boldsymbol{b}\,. \tag{5.1), (5.2}$$

Voraussetzung für die Existenz beider Schreibweisen (1) und (4) ist, daß weder $\boldsymbol{C}$ noch $\boldsymbol{J}$ singulär sind. Es gibt jedoch einige wenige Sonderfälle der Werkstoffmechanik, wo endweder $\boldsymbol{C}$ oder $\boldsymbol{J}$ singulär werden. Auf diese Sonderfälle wird im weiteren noch besonders hingewiesen. Der Vektor $\boldsymbol{h}$ hat im allgemeinen eine Dimension, die sich wesentlich von der von $\boldsymbol{\sigma}$ und $\boldsymbol{\varepsilon}$ unterscheidet, so daß die Matrix $\boldsymbol{A}$ rechteckig ist. Eine Inversion dieser Matrix ist daher nicht möglich, aber auch nicht erforderlich. Lediglich kann unter Umständen die Kombination von (2) und (4) zweckmäßig sein:

$$\boldsymbol{h} = \boldsymbol{B}\boldsymbol{\varepsilon} + \boldsymbol{e} \tag{6}$$

mit

$$\boldsymbol{B} = \boldsymbol{A}\boldsymbol{C} = \boldsymbol{A}\boldsymbol{J}^{-1}, \qquad \boldsymbol{e} = \boldsymbol{c} + \boldsymbol{A}\boldsymbol{d} = \boldsymbol{c} - \boldsymbol{A}\boldsymbol{J}^{-1}\boldsymbol{b}\,. \tag{7.1), (7.2}$$

Einige wenige Eigenschaften des allgemeinen Deformationsgesetzes lassen sich ohne Spezifizierung der Matrizen $\boldsymbol{J}$, $\boldsymbol{A}$, $\boldsymbol{b}$ und $\boldsymbol{c}$ erkennen. Ist entweder $\dot{\boldsymbol{\sigma}}$ oder $\dot{\boldsymbol{\varepsilon}}$ gleich dem Nullvektor, so existiert im allgemeinen der jeweils zugeordnete Vektor $\dot{\boldsymbol{\varepsilon}}$ bzw. $\dot{\boldsymbol{\sigma}}$. Damit ändert sich der Verzerrungszustand auch bei konstantem Spannungszustand (Kriechen), der Spannungszustand gleichfalls bei konstantem Verzerrungszustand (Relaxation). Derartige Erscheinungen charakterisieren allgemein rheonomes Material. Damit dieser Vorgang nicht schon vor jeglicher Erstbelastung oder -verformung einsetzt, müssen die Vektoren $\boldsymbol{b}$ und $\boldsymbol{c}$ verschwinden, wenn alle Zustandsparameter gleich Null sind (wobei die Wahl von $\boldsymbol{h}$ stets so getroffen werden soll, daß dieser Vektor vor der ersten Belastung gleich dem Nullvektor ist):

$$\boldsymbol{b}(\boldsymbol{0}, \boldsymbol{0}, \boldsymbol{0}) = \boldsymbol{0}, \qquad \boldsymbol{c}(\boldsymbol{0}, \boldsymbol{0}, \boldsymbol{0}) = \boldsymbol{0}\,. \tag{8) , (9}$$

Sind $\boldsymbol{b}$ und $\boldsymbol{c}$ für beliebige Zustände identisch den entsprechenden Nullvektoren, so verhält sich das Material skleronom. Änderungen des Verzerrungszustandes ergeben sich nur bei Änderung des Spannungszustandes und umgekehrt. Physikalisch ist dabei das Verschwinden von $\boldsymbol{c}$ eine notwendige Folgerung aus dem Nullsetzen von $\boldsymbol{b}$, da die internen Zustandsparameter als Größen gedeutet werden können, die die vorangegangene Verformungsgeschichte beschreiben. Sie dürfen sich also gleichfalls nicht ändern, wenn der „äußerlich sichtbare" Zustand $\boldsymbol{\sigma}$, $\boldsymbol{\varepsilon}$ konstant bleibt.

Alle weiteren Eigenschaften der Matrizen $\boldsymbol{J}$, $\boldsymbol{A}$, $\boldsymbol{b}$ und $\boldsymbol{c}$ sollen nun an speziellen Verhaltensweisen des Materials diskutiert werden. Als Demonstrationsbeispiel soll der ebene Hauptspannungszustand

$$\boldsymbol{\sigma} = \begin{Bmatrix} \sigma_1 \\ \sigma_2 \end{Bmatrix}. \tag{10}$$

dienen, als Verzerrungsvektor wird entsprechend

$$\boldsymbol{\varepsilon} = \begin{Bmatrix} \varepsilon_1 \\ \varepsilon_2 \end{Bmatrix} \tag{11}$$

definiert. In allen Beispielen wird Isotropie vorausgesetzt, so daß die Hauptdehnungen parallel zu den Hauptspannungen liegen.

3. Material mit skleronomen Eigenschaften

3.1. Elastisches Material

Im elastischen Zustand sind alle Verzerrungskomponenten eindeutige und umkehrbare Funktionen der Spannungskomponenten. Diese Funktionen ergeben sich als partielle Ableitungen einer skalaren Funktion (Potential) nach den zugeordneten Komponenten. Ist dieses Potential die spezifische Ergänzungsarbeit und als Funktion der Kom-

[1]) Dabei soll das Temperaturfeld als gegeben vorausgesetzt werden. Die Berücksichtigung der Rückwirkung der Dissipation auf das Temperaturfeld führt zu wesentlich komplexeren Stoffgesetzen.

ponenten des Spannungstensors formuliert (z. B. [1], S. 56), dann sind die Verzerrungskomponenten partielle Ableitungen der Ergänzungsarbeit nach den zugeordneten Spannungskomponenten. Differentiation dieses Vektors nach der Zeit ergibt sofort den linearen Zusammenhang (1) mit $\boldsymbol{b} = \boldsymbol{0}$. Die Matrix $\boldsymbol{J}(\boldsymbol{\sigma})$ wird aus partiellen Differentialquotienten 2. Ordnung gebildet und ist symmetrisch.

Analog ergibt sich die inverse Matrix $\boldsymbol{C}(\boldsymbol{\varepsilon})$ aus partiellen Ableitungen 2. Ordnung der spezifischen Arbeit nach den Komponenten des Verzerrungstensors. Sie ist in dem Zusammenhang $\boldsymbol{C}(\boldsymbol{\sigma})$ durch Einsetzen von $\boldsymbol{\varepsilon}(\boldsymbol{\sigma})$ überführbar, wenn das nichtlineare Elastizitätsgesetz bequem invertierbar ist. Ist dies nicht der Fall, kann natürlich die Inversion der Matrizen $\boldsymbol{J}$ in $\boldsymbol{C}$ und umgekehrt nach (5.1) durchgeführt werden.

Bereitet die Inversion von $\boldsymbol{\varepsilon}(\boldsymbol{\sigma})$ Mühe, so kann sie in Sonderfällen mit Hilfe einer Hilfsgröße $\boldsymbol{h}$ umgangen werden. Folgendes Beispiel soll das illustrieren. Ein isotropes Elastizitätsgesetz soll die Form $\boldsymbol{\sigma} = \boldsymbol{\sigma}(\boldsymbol{\varepsilon})$ haben. Im ebenen Hauptspannungszustand (10) ist $\sigma_3 = \sigma_{33} = 0$. Aus dem entsprechenden nichtlinearen Zusammenhang zwischen σ_3 und $\varepsilon_1, \varepsilon_2, \varepsilon_3$ muß ε_3 eliminiert werden, um eine quadratische Steifigkeitsmatrix zu erzielen. Das wird im allgemeinen nur numerisch möglich sein. Dieser Auflösung schließt sich dann zwangsläufig eine numerische Differentiation an. Diese beiden Prozesse lassen sich umgehen, wenn vor der Elimination von ε_3 das Elastizitätsgesetz differenziert wird und (mit der für die Zuwüchse linearen Gleichung $\dot{\sigma}_3 = 0$) $\dot{\varepsilon}_3$ aus dem Zusammenhang zwischen (10) und (11) eliminiert wird. Die Elemente von $\boldsymbol{C}$ enthalten dann aber weiterhin ε_3. Wird ε_3 nun formal als Zustandsvariable $\boldsymbol{h} = \{\varepsilon_3\}$ aufgefaßt, so entsteht sofort die Struktur (4), (6) mit einer (1, 2)-Matrix $\boldsymbol{B}$. In jedem Fall verschwinden die Vektoren $\boldsymbol{b}$ und $\boldsymbol{c}$.

Als Demonstrationsbeispiel soll tensoriell lineares isotropes Material mit linearem Verhalten der Volumenänderung dienen [2]:

$$e_{ij} = \frac{1}{2G} g\,(\sigma_V)\, s_{ij}, \qquad e = \frac{1}{3K} s\,. \tag{12.1}., (12.2)$$

Hierin ist G der Schubmodul des linearen Grenzfalls, K der Kompressionsmodul, e_{ij} und s_{ij} sind die Verzerrungs- und Spannungsdeviatoren, e die Normaldehnungssumme und s die Normalspannungssumme. σ_V ist die Vergleichsspannung,

$$\sigma_v = \sqrt{\tfrac{3}{2}\, s_{kl} s_{kl}}\,, \tag{13}$$

die den Wert der 2. Invariante eines beliebigen Spannungszustandes mit dem entsprechenden des einachsigen Zuges vergleicht. Dann folgt für den ebenen Hauptspannungszustand:

$$\boldsymbol{J}(\boldsymbol{\sigma}) = \frac{1}{E}\begin{bmatrix} 1 + \frac{2}{3}(1+\nu)\left(\bar{g} + \bar{g}'\,\frac{(\sigma_1 - \frac{1}{2}\sigma_2)^2}{\sigma_V}\right) & -\nu - \frac{1}{3}(1+\nu)\left(\bar{g} - 2\bar{g}'\,\frac{(\sigma_1 - \frac{1}{2}\sigma_2)(\sigma_2 - \frac{1}{2}\sigma_1)}{\sigma_V}\right) \\ -\nu - \frac{1}{3}(1+\nu)\left(\bar{g} - 2\bar{g}'\,\frac{(\sigma_1 - \frac{1}{2}\sigma_2)(\sigma_2 - \frac{1}{2}\sigma_1)}{\sigma_V}\right) & 1 + \frac{2}{3}(1+\nu)\left(\bar{g} + \bar{g}'\,\frac{(\sigma_2 - \frac{1}{2}\sigma_1)^2}{\sigma_V}\right) \end{bmatrix} \tag{14}$$

mit

$$\bar{g} = g(\sigma_V) - 1, \qquad \bar{g}' = \frac{\mathrm{d}g(\sigma_V)}{\mathrm{d}\sigma_V}, \qquad \sigma_V = \sqrt{\sigma_1^2 - \sigma_1\sigma_2 + \sigma_2^2} \tag{14.1), (14.2), (14.3}$$

und den für Isotropie gültigen Beziehungen zwischen G, K und E, ν (z. B. [3], S. 277).

Liegt für den einachsigen Versuch die Spannungsdehnungskurve $\varepsilon = \varepsilon(\sigma)$ und für kleine Spannungen eine Information über die Querdehnung $\varepsilon_Q = \varepsilon_Q(\sigma)$ vor, so bestimmen sich E, ν und $\bar{g}$ aus

$$E = \left(\left.\frac{\mathrm{d}\varepsilon(\sigma)}{\mathrm{d}\sigma}\right|_{\sigma=0}\right)^{-1}, \qquad \nu = -E\left.\frac{\mathrm{d}\varepsilon_Q(\sigma)}{\mathrm{d}\sigma}\right|_{\sigma=0}, \qquad \bar{g} = \left.\frac{\varepsilon(\sigma) - 1/E\cdot\sigma}{2/3\,(1+\nu)\,\sigma}\right|_{\sigma=\sigma_V}. \tag{15.1), (15.2), (15.3}$$

Im linearen Fall ist $\bar{g} = 0$ und die Matrix $\boldsymbol{J}$ hängt nicht mehr von $\boldsymbol{\sigma}$ ab:

$$\boldsymbol{J} = \frac{1}{E}\begin{bmatrix} 1 & -\nu \\ -\nu & 1 \end{bmatrix}. \tag{16}$$

Hieraus folgt dann

$$\boldsymbol{C} = \frac{E}{1-\nu^2}\begin{bmatrix} 1 & \nu \\ \nu & 1 \end{bmatrix}. \tag{17}$$

Für alle technisch sinnvollen Fälle sind sowohl $\boldsymbol{J}$ als auch $\boldsymbol{C}$ im ebenen Spannungszustand stets regulär.[1]) Das gilt auch für alle anderen Spannungszustände, solange elastisches Material kompressibel ist. Inkompressibles Material (d. h., $e = 0$ im Falle kleiner Verzerrungen) liefert singuläre $\boldsymbol{J}$-Matrizen dann, wenn $\boldsymbol{\varepsilon}$ alle 3 Normaldehnungen enthält bzw. wenn von einer Normaldehnung $\varepsilon_3 = 0$ gefordert wird (ebener Verzerrungszustand). In diesen Fällen muß die Lösung des Randwertproblems, sofern sie die Matrix $\boldsymbol{C}$ erfordert, modifiziert werden, da der seitens des Deformationsgesetzes unbestimmte Wert von s nur durch die Gleichgewichtsbedingungen und die statischen Randbedingungen bestimmt werden kann.

3.2. Elastisch-plastisches Material

Wir betrachten zunächst isotropes elastisch-plastisches Material mit isotroper Verfestigung mit der Fließbedingung von HUBER-MISES. Das Deformationsgesetz hat folgende Gestalt, wenn die Erfüllung des Postulats von DRUCKER

[1]) Aus Arbeitsbetrachtungen folgt für lineares Materialverhalten $G \geqq 0$, $K \geqq 0$. Der Grenzfall $G = 0$, $K > 0$ ist deshalb bei phänomenologischer Betrachtung nicht auszuschließen. Hieraus folgt aber wegen $\nu = -1$ eine singuläre Matrix $\boldsymbol{J}$. Der technisch sinnvolle Bereich ist dagegen durch $0 \leqq \nu \leqq 0{,}5$ charakterisiert.

vorausgesetzt wird (z. B. [4], [5]):

$$\dot e_{ij} = \frac{1}{2G}\dot s_{ij} + \dot e_{ij}^P\,, \qquad \dot e = \frac{1}{3K}\dot s \tag{18.1), (18.2}$$

mit

$$\dot e_{ij}^P = \begin{cases} 0 & \text{wenn} \quad \sigma_V < k(\varepsilon_V^P) \quad \text{oder} \quad \sigma_V = k,\ \dot\sigma_V > 0 \\ \dot\lambda s_{ij}\,, & \phantom{\text{wenn}} \quad \sigma_V = k(\varepsilon_V^P) \quad \text{und} \quad \dot\sigma_V \geqq 0 \end{cases} \tag{19}$$

und

$$\dot\varepsilon_V^P = \sqrt{\tfrac{2}{3}\,\dot e_{kl}^P \dot e_{kl}^P}\,. \tag{20}$$

Wird vorausgesetzt, daß die Spannungs-Dehnungskurve bei Erstbelastung wieder durch $\varepsilon = \varepsilon_E(\sigma)$ beschrieben wird, ist $k(\varepsilon_V^P)$ die Umkehrfunktion zu

$$\varepsilon^P\Big|_{\varepsilon^P = \varepsilon_V^P} = \left[\varepsilon_E(\sigma) - \frac{1}{E}\sigma\right]\Big|_{\sigma = k}. \tag{21}$$

Der Parameter $\dot\lambda$ kann aus Fließbedingung und Fließgesetz in den beiden Formen

$$\dot\lambda = \alpha_{kl}\dot\varepsilon_{kl} = \frac{3}{2}\,\frac{\sigma_{kl} - \frac{1}{3}\sigma_{mm}\delta_{kl}}{k^2(1 + k'/3G)}\,\dot\varepsilon_{kl} \tag{22}$$

bzw.

$$\dot\lambda = \beta_{kl}\dot\sigma_{kl} = \frac{9}{4}\,\frac{\sigma_{kl} - \frac{1}{3}\sigma_{mm}\delta_{kl}}{k^2 k'}\,\dot\sigma_{kl} \tag{23}$$

bestimmt werden. Es ist sofort ersichtlich, daß $\dot\lambda$ nur im Falle der Verfestigung durch $\dot\sigma_{kl}$ ausgedrückt werden kann, weil

$$k' = \frac{\mathrm{d}k(\varepsilon_V^P)}{\mathrm{d}\varepsilon_V^P} \tag{24}$$

gegen Null geht, wenn sich das Material nicht oder nicht mehr verfestigt. Das bedeutet, daß die Faktoren von $\dot\sigma_{kl}$ gegen unendlich streben, bei ideal-plastischem Material sofort bei Fließbeginn, bei verfestigendem dann, wenn die Kurve k sich an eine Asymptote $k_\infty = \text{const.}$ annähert. Hierin drückt sich formal der bekannte Umstand aus, daß das Fließen bei ideal plastischem Material unbegrenzt ist. Dagegen liefert (22) stets eindeutige Werte von $\dot\lambda$. Der einzige Zustandsparameter ist

$$\boldsymbol{h} = \{\varepsilon_V^P\}\,. \tag{25}$$

Für den ebenen Hauptspannungszustand gibt (18), (19) mit (10), (11)

$$\boldsymbol{J}(\boldsymbol{\sigma}, \boldsymbol{h}) = \frac{1}{E}\begin{bmatrix} 1 + p\dfrac{E}{k^2k'}\left(\sigma_1 - \dfrac{1}{2}\sigma_2\right)^2 & -\nu + p\dfrac{E}{k^2k'}\left(\sigma_1 - \dfrac{1}{2}\sigma_2\right)\left(\sigma_2 - \dfrac{1}{2}\sigma_1\right) \\ -\nu + p\dfrac{E}{k^2k'}\left(\sigma_1 - \dfrac{1}{2}\sigma_2\right)\left(\sigma_2 - \dfrac{1}{2}\sigma_1\right) & 1 + p\dfrac{E}{k^2k'}\left(\sigma_2 - \dfrac{1}{2}\sigma_1\right)^2 \end{bmatrix} \tag{26}$$

mit

$$p = \begin{cases} 1 & \text{wenn } \sigma_V = k(\varepsilon_V^P) \text{ und } \dot\sigma_V \geqq 0\,, \\ 0 & \text{sonst}\,, \end{cases}$$

und

$$\boldsymbol{A}(\boldsymbol{\sigma}, \boldsymbol{h}) = p\left[\frac{\sigma_1 - \frac{1}{2}\sigma_2}{kk'} \quad \frac{\sigma_2 - \frac{1}{2}\sigma_1}{kk'}\right]. \tag{27}$$

Da elastisch-plastisches Material sich skleronom verhält, sind die Vektoren $\boldsymbol{b}$ und $\boldsymbol{c}$ wiederum Nullvektoren. Die Inverse zu (26) lautet

$$\boldsymbol{C}(\boldsymbol{\sigma}, \boldsymbol{h}) = \frac{E}{N}\begin{bmatrix} \dfrac{k^2k'}{E} + p\left(\sigma_2 - \dfrac{1}{2}\sigma_1\right)^2 & \nu\dfrac{k^2k'}{E} - p\left(\sigma_1 - \dfrac{1}{2}\sigma_2\right)\left(\sigma_2 - \dfrac{1}{2}\sigma_1\right) \\ \nu\dfrac{k^2k'}{E} - p\left(\sigma_1 - \dfrac{1}{2}\sigma_2\right)\left(\sigma_2 - \dfrac{1}{2}\sigma_1\right) & \dfrac{k^2k'}{E} + p\left(\sigma_1 - \dfrac{1}{2}\sigma_2\right)^2 \end{bmatrix} \tag{28}$$

mit

$$N = \frac{k^2k'}{E}(1 - \nu^2) + p\left[\left(\sigma_1 - \frac{1}{2}\sigma_2\right)^2 + \left(\sigma_2 - \frac{1}{2}\sigma_1\right)^2 + 2\nu\left(\sigma_1 - \frac{1}{2}\sigma_2\right)\left(\sigma_2 - \frac{1}{2}\sigma_1\right)\right]. \tag{28.1}$$

Gemäß (7.1) ergibt sich

$$\boldsymbol{B}(\boldsymbol{\sigma}, \boldsymbol{h}) = \frac{pk}{N}\left[\left[\left(\sigma_1 - \frac{1}{2}\sigma_2\right) + \nu\left(\sigma_2 - \frac{1}{2}\sigma_1\right)\right]\left[\left(\sigma_2 - \frac{1}{2}\sigma_1\right) + \nu\left(\sigma_1 - \frac{1}{2}\sigma_2\right)\right]\right]. \tag{29}$$

Für $k' = 0$ bleibt $\boldsymbol{C}$ beschränkt, wird aber singulär. Daher entartet $\boldsymbol{J}$. Auch $\boldsymbol{A}$ geht gegen unendlich, während $\boldsymbol{B}$ existiert. Das ist genau der Effekt, der bereits für den allgemeinsten Fall im Zusammenhang mit $\dot\lambda$ erläutert worden ist. Für die Lösung von Randwertaufgaben, die von Verformungsansätzen ausgehen (z. B. Theorie der Flächentragwerke, Deformationsmethode bei finiten Elementen) werden aber gerade die Darstellungen (4) und (6) mit (28) und (29) benötigt. Daher kann auch der ideal-plastische Fall ohne weiteres eingeschlossen werden, wobei noch die Vereinfachung eintritt, daß weder $\boldsymbol{h}$ noch $\boldsymbol{B}$ benötigt werden und sich das Deformationsgesetz auf

$$\dot{\boldsymbol{\sigma}} = \boldsymbol{C}(\boldsymbol{\sigma})\,\dot{\boldsymbol{\varepsilon}} \tag{30}$$

reduziert.

Etwas aufwendiger bezüglich der Größe $\boldsymbol{h}$ ist der Fall kinematischer Verfestigung. Für den Fall, daß $\varepsilon = \varepsilon_E(\sigma)$ wiederum die allgemeine nichtlineare Spannungs-Dehnungsfunktion im einachsigen Erstbelastungsversuch ist, hat Iwan [6] ein Modell vorgestellt, welches das lineare kinematische Verfestigungsgesetz von Prager [5] durch eine kontinuierliche Verteilung von unendlich vielen Fließflächen mit einer der jeweiligen plastischen Verzerrung proportionalen Verschiebung des Mittelpunktes ersetzt:

$$\dot{e}_{ij} = \frac{1}{2G}\dot{s}_{ij} + \int_0^k \dot{\alpha}_{ij}(\bar{k})\,\Gamma(\bar{k})\,\mathrm{d}\bar{k}, \qquad \dot{e} = \frac{1}{3K}\dot{s} \tag{31.1), (31.2}$$

mit

$$\dot{\alpha}_{ij}(k) = \begin{cases} \dfrac{3}{2k^2}\big(s_{ij} - \alpha_{ij}(k)\big)\big(s_{kl} - \alpha_{kl}(k)\big)\,\dot{s}_{kl} & \text{wenn } (s_{kl} - \alpha_{kl})\,(s_{kl} - \alpha_{kl}) = \dfrac{2}{3}k^2 \\ & \text{und } (s_{kl} - \alpha_{kl})\,\dot{s}_{kl} > 0\,, \\ 0 & \text{sonst.} \end{cases} \tag{32}$$

Die Funktion $\Gamma(k)$ bestimmt sich aus der Erstbelastungskurve $\varepsilon = \varepsilon_E(\sigma)$ zu

$$\Gamma(k) = \frac{2}{3}\left.\frac{\mathrm{d}^2\varepsilon_E(\sigma)}{\mathrm{d}\sigma^2}\right|_{\sigma=k}. \tag{33}$$

Die Integration über k in (31.1) muß numerisch erfolgen. Dazu wird die Erstbelastungskurve vom Fließbeginn bei k_0 bis zum höchsten Wert der Messung $k_{\max}$ in N (hier zur Vereinfachung gleich groß angenommene) Intervalle

$$\Delta k = \frac{k_{\max} - k_0}{N} \tag{34}$$

zerlegt, so daß $N + 1$ Stützstellen

$$k_n = k_0 + n\Delta k, \qquad 0 \leqq n \leqq N \tag{35}$$

mit $N + 1$ Funktionswerten $\Gamma_n = \Gamma(k_n)$ vorliegen. Für jede dieser Stützstellen wird die Lage des Mittelpunktes $\alpha_{ij}^n = \alpha_{ij}(k_n)$ der Fließfläche durch numerische Integration über die Zeit t gemäß (32) verfolgt. Es gilt dann

$$\int_0^k \dot{\alpha}_{ij}(\bar{k})\,\Gamma(\bar{k})\,\mathrm{d}\bar{k} = \sum_{n=0}^{N} g_n\dot{\alpha}_{ij}^n\Gamma_n\Delta k\,, \tag{36}$$

wobei die g_n die Integrationsgewichte des entsprechenden Verfahrens der Quadratur sind. Mit diesen Voraussetzungen läßt sich nun die Standardform des Deformationsgesetzes für den ebenen Hauptspannungszustand entwickeln. Hierbei müssen in dem Vektor $\boldsymbol{h}$ alle erforderlichen Werte der α_{ij}^n untergebracht werden. Mit

$$\alpha_1^n = \alpha_{11}^n, \qquad \alpha_2^n = \alpha_{22}^n \tag{37}$$

soll $\boldsymbol{h}$ den Aufbau

$$\boldsymbol{h}^T = \{\alpha_1^0, \alpha_2^0, \alpha_1^1, \alpha_2^1, \dots, \alpha_1^N, \alpha_2^N\} \tag{38}$$

haben. Die Dimension beträgt also $2(N + 1)$ und hängt damit von der Zahl der Stützstellen der Integration über k ab.

Es ergeben sich

$$\boldsymbol{J}(\boldsymbol{\sigma}, \boldsymbol{h}) = \frac{1}{E}\begin{bmatrix} 1 + E\sum\limits_{n=0}^{N} g_nA_{11}^n\Gamma_n\Delta k & -\nu + E\sum\limits_{n=0}^{N} g_nA_{12}^n\Gamma_n\Delta k \\ -\nu + E\sum\limits_{n=0}^{N} g_nA_{21}^n\Gamma_n\,\Delta k & 1 + E\sum\limits_{n=0}^{N} g_nA_{22}^n\Gamma_n\,\Delta k \end{bmatrix}, \tag{39}$$

$$\boldsymbol{A}(\boldsymbol{\sigma}, \boldsymbol{h}) = \begin{bmatrix} \boldsymbol{A}^{(0)} \\ \boldsymbol{A}^{(1)} \\ \dots \\ \boldsymbol{A}^{(N)} \end{bmatrix} \tag{40}$$

mit

$$\boldsymbol{A}^{(n)}(\boldsymbol{\sigma}, \boldsymbol{h}) = p\begin{bmatrix} \dfrac{2}{3k_n^2}\left(\sigma_1 - \dfrac{1}{2}\sigma_2 - \dfrac{3}{2}\alpha_1^n\right)^2 & \dfrac{2}{3k_n^2}\left(\sigma_1 - \dfrac{1}{2}\sigma_2 - \dfrac{3}{2}\alpha_1^n\right)\left(\sigma_2 - \dfrac{1}{2}\sigma_1 - \dfrac{3}{2}\alpha_2^n\right) \\ \dfrac{2}{3k_n^2}\left(\sigma_1 - \dfrac{1}{2}\sigma_2 - \dfrac{3}{2}\alpha_1^n\right)\left(\sigma_2 - \dfrac{1}{2}\sigma_1 - \dfrac{3}{2}\alpha_2^n\right) & \dfrac{2}{3k_n^2}\left(\sigma_2 - \dfrac{1}{2}\sigma_1 - \dfrac{3}{2}\alpha_2^n\right)^2 \end{bmatrix} \tag{41}$$

$$p = \begin{cases} 1, \text{ wenn } & \sigma_1^2 - \sigma_1\sigma_2 + \sigma_2^2 - 3(\alpha_1^n\sigma_1 + \alpha_2^n\sigma_2) + 3\big((\alpha_1^n)^2 + \alpha_1^n\alpha_2^n + (\alpha_2^n)^2\big) = k_n^2 \\ \quad\text{und} & (\sigma_1 - \frac{1}{2}\,\sigma_2 - \frac{3}{2}\alpha_1^n)\,\dot{\sigma}_1 + (\sigma_2 - \frac{1}{2}\,\sigma_1 - \frac{3}{2}\,\alpha_2^n)\,\dot{\sigma}_2 > 0\,. \\ 0 \text{ sonst.} \end{cases}$$

Bei kinematischen Modellen wird der Übergang zum idealplastischen Fall in der Regel ausgeschlossen, so daß die Nachgiebigkeitsmatrix $\boldsymbol{J}$ regulär ist und die Inverse $\boldsymbol{C}$ sowie die Matrix $\boldsymbol{B}$ mühelos numerisch berechnet werden können.

Wird versucht, anstelle des Modells (31), (32) mit einer kontinuierlichen Verteilung unendlich vieler Fließflächen auf eine einzige überzugehen, so läßt sich der beobachtete Bauschinger-Effekt nur realistisch beschreiben. wenn für den Weg des Mittelpunktes der Fließfläche ein Relaxationsintegral angesetzt wird [5], [7]. Backhaus schlägt dafür den Ansatz

$$\alpha_{ij} = \int_0^{\varepsilon_V^P} \frac{\mathrm{d}}{\mathrm{d}\bar{\varepsilon}_V^P} [c(\bar{\varepsilon}_V^P)\, \varphi(\varepsilon_V^P - \bar{\varepsilon}_V^P)] \frac{\mathrm{d}\varepsilon_{ij}^P(\bar{\varepsilon}_V^P)}{\mathrm{d}\bar{\varepsilon}_V^P} \mathrm{d}\bar{\varepsilon}_V^P \tag{42}$$

mit ε_V^P nach (20) vor. Ein derartiger Ansatz ist zunächst nicht in den Rahmen (1), (2) einpaßbar. Eine Analogiebetrachtung zur Viskoelastizitätstheorie [8], [9] zeigt aber, daß hier die spezielle Wahl der Funktion φ darüber entscheidet, ob (42) in einen Satz von Differentialgleichungen umwandelbar ist oder nicht.

In [7] wird

$$\varphi(\varepsilon_V^P - \bar{\varepsilon}_V^P) = \varphi_0 + \varphi_1 \, \mathrm{e}^{-\varkappa(\varepsilon_V^P - \bar{\varepsilon}_V^P)\varrho} \tag{43}$$

vorgeschlagen. Dieser Ansatz erfüllt die Bedingung der Überführbarkeit nur im Sonderfall $\varrho = 1$. Nun zeigen aber Untersuchungen in der Viskoelastizitätstheorie, daß Funktionen des Charakters entsprechend (43) stets auch durch eine Exponentialreihe

$$\varphi(\varepsilon_V^P - \bar{\varepsilon}_V^P) = \varphi_0 + \sum_{n=1}^{N} \varphi_n \, \mathrm{e}^{-\varkappa_n(\varepsilon_V^P - \bar{\varepsilon}_V^P)} \tag{44}$$

wenigstens mit der Qualität der Meßwertgenauigkeit approximiert werden können [9]. Mit diesem Ansatz ist (42) in folgender Weise zerlegbar:

$$\alpha_{ij} = \sum_{n=0}^{N} \alpha_{ij}^n , \tag{45.1}$$

$$\alpha_{ij}^n = \int_0^{\varepsilon_V^P} \left(\varphi_n[c'(\bar{\varepsilon}_V^P) + \varkappa_n c \bar{\varepsilon}_V^P)]\, \mathrm{e}^{-\varkappa_n(\varepsilon_V^P - \bar{\varepsilon}_V^P)}\right) \frac{\mathrm{d}\varepsilon_{ij}^P}{\mathrm{d}\bar{\varepsilon}_V^P} \mathrm{d}\bar{\varepsilon}_V^P \quad \text{mit} \quad c'(\varepsilon_V^P) = \frac{\mathrm{d}c(\varepsilon_V^P)}{\mathrm{d}\varepsilon_V^P}, \qquad \varkappa_0 = 0 . \tag{45.2}$$

Durch Differentiation lassen sich die Integrale nun leicht eliminieren, und es folgt

$$\dot{\alpha}_{ij}^n = \varphi_n[c'(\varepsilon_V^P) + \varkappa_n c(\varepsilon_V^P)]\, \dot{\varepsilon}_{ij}^P - \varkappa_n \alpha_{ij}^n \dot{\varepsilon}_V^P . \tag{46}$$

Wird in der Fließbedingung

$$F = \tfrac{1}{2}\,(s_{kl} - \alpha_{kl})\,(s_{kl} - \alpha_{kl}) - \tfrac{1}{3}\,k^2(\varepsilon_V^P) = 0 \tag{47}$$

sowohl die kinematische als auch die isotrope Verfestigung berücksichtigt, so folgt aus dieser Bedingung und dem assozierten Fließgesetz

$$\dot{\varepsilon}_{ij}^P = \begin{cases} \dot{\lambda}(s_{ij} - \alpha_{ij}) & \text{wenn } F = 0 \text{ und } (s_{kl} - \alpha_{kl})\, \dot{s}_{kl} > 0 , \\ 0 & \text{sonst} \end{cases} \tag{48}$$

im Verfestigungsbereich

$$\dot{\lambda} = \frac{3}{2k} \cdot \frac{\sigma_{kl} - \frac{1}{3}\,\sigma_{mm}\delta_{kl} - \alpha_{kl}}{\frac{2}{3}\,kk' + \sum_{n=0}^{N} \varphi_n(c' + \varkappa_n c) - \varkappa_n \alpha_{rs}^n \left(\sigma_{rs} - \frac{1}{3}\,\sigma_{mm}\delta_{rs} - \alpha_{rs}\right)}\, \dot{\sigma}_{kl} . \tag{49}$$

Mit (20), (45), (46), (47), (48) und (49) liegt der vollständige Formelsatz vor, um alle $\dot{\varepsilon}_{ij}^P$, $\dot{\alpha}_{ij}^0$ bis $\dot{\alpha}_{ij}^N$ und $\dot{\varepsilon}_V^P$ als Linearkombinationen von $\dot{\sigma}_{kl}$ (bzw. $\dot{s}_{kl}$) auszudrücken. Somit ist der Zugang zu (1), (2) gewiesen.

Im ebenen Hauptspannungszustand enthält $\boldsymbol{h}$ genau die Elemente wie im vorangegangenen Beispiel nach (38) zuzüglich des Elementes (25) infolge der gleichzeitigen Berücksichtigung der isotropen Verfestigung. Daher ist $\boldsymbol{J}(\boldsymbol{\sigma}, \boldsymbol{h})$ eine (2,2)-Matrix, während $\boldsymbol{A}(\boldsymbol{\sigma}, \boldsymbol{h})$ $2N + 3$ Zeilen und zwei Spalten hat. Im Gegensatz zum Modell (31) repräsentiert N nicht die frei wählbare Stützstellenzahl, sondern einen Optimalwert aus der Approximation, die nach in [9] ausführlich diskutierten Methoden erfolgen kann. Auf eine explizite Angabe von $\boldsymbol{J}$ und $\boldsymbol{A}$ soll verzichtet werden.

4. Material mit rheonomen Eigenschaften

4.1. Viskoelastisches Material

Viskoelastisches Material wird von viskoplastischem entsprechend einer Definition, wie sie in [10] vorgeschlagen ist, unterschieden. Mit den Arbeiten von Green und Rivlin [11] ist für derartiges Material erstmals die allgemeinste Form des Deformationsgesetzes vorgelegt worden. Iljuschin und Pobjedrja [3] haben dieses Konzept durch Untersuchung der Sonderfallstruktur weitergeführt. Im allgemeinsten Fall wird der Zusammenhang zwischen Verzerrungs- und Spannungstensor durch Mehrfachintegraloperatoren beschrieben, denen unter Berücksichtigung von

Singularitäten der Kerne vom Typ der DIRAC-Funktion beliebige Einfachintegraloperatoren und Differentialoperatoren als Sonderfälle zugeordnet werden können. Die allgemeine Theorie ist nicht in die Form (1), (2) zu bringen. Voraussetzung dafür ist der Sonderfall Differentialoperator. Die Möglichkeit eines systematischen Zuganges zu diesem Sonderfall aus der allgemeinen Theorie besteht über folgende zwei Approximationen:

1. Der allgemeine Mehrfachintegraloperator wird durch seinen Hauptteil (in der Definition des Hauptteiles nach [3], S. 172) angenähert.
2. Der Kern des Hauptteils wird durch eine Summe von Exponentialfunktionen (Exponentialreihe) approximiert.

Die erste Approximation ist nur erforderlich, wenn ein nichtlinearer Spannungs-Verzerrungs-Zusammenhang vorliegt, da die lineare Theorie ausschließlich aus dem Hauptteil (Einfachintegraloperator) besteht. In der Arbeit [9] wird an Hand von Analysen einer Reihe von Veröffentlichungen die Brauchbarkeit dieser Approximation für die erforderliche und mögliche Genauigkeit bestätigt.

Die zweite Approximation ist auf eine reine Fehlerbetrachtung zwischen den direkt meßbaren Werten des Operatorkerns und den entsprechenden Werten einer Ansatzfunktion zu reduzieren. Die Schwierigkeit besteht hier nur im Verfahren der Parameterermittlung, da es sich hier um ein nichtlineares Optimierungsproblem handelt. Praktikable Näherungsmethoden sind in [9] zusammengestellt, so daß diese Approximation prinzipiell keine Schwierigkeiten bereitet. Erfahrungen mehrerer Jahre an vielen praktischen Beispielen bestätigen, daß die Fehler der Approximation im Bereich der Genauigkeit der experimentellen Daten liegen.

Im linearen Grenzfall ist die zweite Approximation eindeutig für die Abbildung des Integraloperators durch einen Differentialoperator. Die nichtlineare Erweiterung ist jedoch auch für andere Differentialoperatoren, die nicht nur den Hauptteil enthalten, möglich. Beispiele dafür sind in [2], [9] zu finden.

Eine Standardform eines isotropen Deformationsgesetzes [2], [8], [12], welches beide Approximationen erfüllt, soll als Beispiel zum Aufbau des Deformationsgesetzes (1) vorgestellt werden:

$$e_{ij}(t) = \frac{1}{2G}\left[g(\sigma_V(t))\, s_{ij}(t) + \int_{-\infty}^{t} C(t-\tau,\ \sigma_V(\tau))\, s_{ij}(\tau)\, d\tau\right], \tag{50.1}$$

$$e(t) = \frac{1}{3K}\, s(t) \tag{50.2}$$

mit

$$C(t, \sigma_V) = H(t)\left[h(\sigma_V)\frac{1}{T} + f(\sigma_V)\sum_{n=1}^{N}\frac{A_n}{T_n}\, e^{\left(-\frac{t}{T_n}\right)}\right]. \tag{50.3}$$

Hierin sind G, K, e_{ij}, e, s_{ij}, s und σ_V die Größen, die beim elastischen Gesetz erläutert worden sind. Die Funktion $H(t)$ ist die HEAVISIDE-Funktion, h und f sind Funktionen ähnlichen Charakters wie g und A_n, T_n Konstanten aus der Annäherung der Kriechkurven durch die Exponentialreihe. Ausführlichere Darstellungen zur Ermittlung aller Parameter finden sich in [12]. Die Hilfsgrößen

$$h_{ij}^n(t) = \frac{A_n}{2GT_n}\int_{-\infty}^{t} e^{\left(-\frac{t-\tau}{T_n}\right)} f(\sigma_V(\tau))\, s_{ij}(\tau)\, d\tau, \tag{51}$$

die nach Einsetzen von (50.3) in (50.1) Teile des Integraloperators repräsentieren, müssen den Differentialgleichungen

$$\dot{h}_{ij}^n = \frac{1}{T_n}\left(\frac{A_n}{2G} f(\sigma_V)\, s_{ij} - h_{ij}^n\right) \tag{52}$$

genügen, die unmittelbar aus der differenzierten Gleichung (51) und Ersatz des Integraloperators mit Hilfe von (51) folgen. Werden (50.1) und (50.2) gleichfalls nach t differenziert, entsteht zusammen mit (52) ein Satz simultaner Differentialgleichungen. Im ebenen Hauptspannungszustand ergeben sich mit

$$h_1^n = h_{11}^n, \quad h_2^n = h_{22}^n, \qquad \boldsymbol{h}^T = \{h_1^1, h_2^1, h_1^2, h_2^2, \dots, h_1^N, h_2^N\} \tag{53), (54}$$

folgende Teilmatrizen des Gesetzes (1), (2):

$$\boldsymbol{A} = \boldsymbol{0}, \tag{55}$$

$$\boldsymbol{b}(\boldsymbol{\sigma}, \boldsymbol{h}) = \begin{Bmatrix} \frac{2}{3}\frac{1+\nu}{ET} h(\sigma_V)\left(\sigma_1 - \frac{1}{2}\sigma_2\right) + \sum_{n=1}^{N} c_1^n(\boldsymbol{\sigma}, \boldsymbol{h}) \\ \frac{2}{3}\frac{1+\nu}{ET} h(\sigma_V)\left(\sigma_2 - \frac{1}{2}\sigma_1\right) + \sum_{n=1}^{N} c_2^n(\boldsymbol{\sigma}, \boldsymbol{h}) \end{Bmatrix}, \tag{56}$$

$$\boldsymbol{c}(\boldsymbol{\sigma}, \boldsymbol{h}) = \begin{Bmatrix} \boldsymbol{c}^{(1)}(\boldsymbol{\sigma}, \boldsymbol{h}) \\ \boldsymbol{c}^{(2)}(\boldsymbol{\sigma}, \boldsymbol{h}) \\ \cdots \\ \boldsymbol{c}^{(N)}(\boldsymbol{\sigma}, \boldsymbol{h}) \end{Bmatrix} \tag{57}$$

$$\boldsymbol{c}^{(n)}(\boldsymbol{\sigma}, \boldsymbol{h}) = \begin{Bmatrix} c_1^n \\ c_2^n \end{Bmatrix} = \frac{1}{T_n}\begin{Bmatrix} \frac{2}{3}\frac{1+\nu}{E} A_n f(\sigma_V)\left(\sigma_1 - \frac{1}{2}\sigma_2\right) - h_1^n \\ \frac{2}{3}\frac{1+\nu}{E} A_n f(\sigma_V)\left(\sigma_2 - \frac{1}{2}\sigma_1\right) - h_2^n \end{Bmatrix} = \frac{1}{T_n}\begin{Bmatrix} f_1^n - h_1^n \\ f_2^n - h_2^n \end{Bmatrix} = \frac{1}{T_n}\begin{Bmatrix} D_1^n \\ D_2^n \end{Bmatrix}. \tag{58}$$

Die Matrix $\boldsymbol{J}$ hängt nur von $\boldsymbol{\sigma}$ ab und entspricht genau (14), da der elastische Teil des Gesetzes (50) mit (12) übereinstimmt. Damit gilt auch das dort Gesagte über die Regularität von $\boldsymbol{J}$, so daß keine numerischen Schwierigkeiten bei der Umformung von (1) in (4) entstehen.

Obwohl mit (14), (55) bis (58) das Deformationsgesetz (1), (2) dem ebenen Spannungszustand von (50) völlig äquivalent ist, sind mit der Umformung von (50) in (1), (2) noch nicht alle Probleme der numerischen Integration gelöst. Eine für diese Materialklasse typische Schwierigkeit resultiert aus der Approximation der realen Kriechkurve durch die Exponentialreihe, die zu sehr unterschiedlichen Größenordnungen der Parameter T_n führt. Wird (2) unter den Bedingungen eines vorgegebenen Spannungszustandes integriert, sind die Differentialgleichungen für alle h_1^n und h_2^n entkoppelt. Die Schrittweite der Integration nach RUNGE-KUTTA läßt sich dann nach [13], S. 417 zu

$$\Delta t = \gamma T_n, \qquad (\gamma \approx 0{,}1 \dots 0{,}2) \tag{59}$$

abschätzen. Bei einer simultanen Integration bestimmt also das kleinste T_n die Schrittweite, obwohl sich die zugeordneten h_1^n und h_2^n bei langsamen Änderungen des Spannungszustandes asymptotisch an die Störfunktionen f_1^n und f_2^n annähern. Diese asymptotische Annäherung kann aber gut genutzt werden, um die Schrittweite in diesem Fall von T_n unabhängig zu machen. Das erfordert folgende Modifikation von (14), (55) bis (58), [2], [8], [9]:

$$\boldsymbol{J} = \boldsymbol{J}_E + \sum_{n=1}^{N} \boldsymbol{A}^{(n)}, \qquad \boldsymbol{A} = \begin{bmatrix} \boldsymbol{A}^{(1)} \\ \boldsymbol{A}^{(2)} \\ \dots \\ \boldsymbol{A}^{(N)} \end{bmatrix}, \qquad (60), \quad (61)$$

$$\boldsymbol{A}^{(n)} = p_n \frac{2}{3}(1+\nu)\frac{A_n}{E} \times$$

$$\times \begin{bmatrix} f(\sigma_V) + \frac{1}{\sigma_V}\left(\sigma_1 - \frac{1}{2}\sigma_2\right)^2 f'(\sigma_V) & -\frac{1}{2} f(\sigma_V) + \frac{1}{\sigma_V}\left(\sigma_1 - \frac{1}{2}\sigma_2\right)\left(\sigma_2 - \frac{1}{2}\sigma_1\right) f'(\sigma_V) \\ -\frac{1}{2} f(\sigma_V) + \frac{1}{\sigma_V}\left(\sigma_1 - \frac{1}{2}\sigma_2\right)\left(\sigma_2 - \frac{1}{2}\sigma_1\right) f'(\sigma_V) & f(\sigma_V) + \frac{1}{\sigma_V}\left(\sigma_2 - \frac{1}{2}\sigma_1\right)^2 f'(\sigma_V) \end{bmatrix} \tag{62}$$

$$\boldsymbol{c}^{(n)} = (1 - p_n)\frac{1}{T_n}\begin{Bmatrix} D_1^n \\ D_2^n \end{Bmatrix}, \tag{63}$$

$$p_n = \begin{cases} 1 & \text{wenn } \|\dot{\boldsymbol{h}}^{(n)}\| \ll \frac{1}{T_n}\|\boldsymbol{h}^{(n)}\| \text{ und } \|\dot{\boldsymbol{f}}^{(n)}\| \ll \frac{1}{T_n}\|\boldsymbol{f}^{(n)}\| \\ 0 & \text{sonst.} \end{cases} \tag{64}$$

$\boldsymbol{J}_E$ ist die ursprüngliche Matrix $\boldsymbol{J}$ nach (14); (56) und (57) bleiben unverändert. Beim Programmieren müssen die Bedingungen zu (64) natürlich als Fehlerschranken quantifiziert werden. Sie sollten dabei weitgehend den Schranken des Integrationsverfahrens entsprechen, ihre Parameter jedoch unterschiedliche Werte für den Eintritt in den und den Austritt aus dem asymptotischen Zustand annehmen [8]. Einzelheiten sind in [9] ausgeführt. Zum Anliegen dieser Arbeit soll nur hervorgehoben werden, daß diese dem Problem angepaßte Methode zur Überwindung der sogenannten Steifigkeit des Systems von Differentialgleichungen (1), (2) eine Matrix $\boldsymbol{A}$ erfordert, die im ursprünglichen System nicht auftritt. Das Auftreten von Elementen in $\boldsymbol{A}$ und das Verschwinden von Elementen zugeordneter Zeilen in $\boldsymbol{c}$ sowie der umgekehrte Vorgang sind vom erreichten Zustand abhängig, wobei auch $\boldsymbol{J}$ geändert wird. $\boldsymbol{J}$ bleibt wegen des konstanten Anteils unabhängig von den Modifikationen regulär.

4.2. Viskoplastisches Material

Im Sinne eines typischen Beispiels sei ein isotroper elastisch-viskoplastischer Werkstoff mit der Fließbedingung von HUBER-MISES und isotroper Verfestigung aus der großen Klasse viskoplastischer Materialien vorgestellt [14], S. 190:

$$\dot{e}_{ij} = \frac{1}{2G}\left[\dot{s}_{ij} + \frac{k_1}{T}\left\langle H\left(\frac{\sigma_V}{\varkappa(\eta)} - k_0\right)\right\rangle \frac{s_{ij}}{\sigma_V}\right], \qquad \dot{e} = \frac{1}{3K}\dot{s}\,. \qquad (65.1), \quad (65.2)$$

Hierin sind k_0 und k_1 Konstanten mit der Dimension einer Spannung, T eine Konstante mit der Dimension der Zeit, $\varkappa$ eine dimensionslose Funktion des Verfestigungsparameters, der wahlweise durch den viskoplastischen Verformungsweg

$$\dot{\eta} = \dot{\varepsilon}_V^{VP} = \sqrt{\tfrac{2}{3}\,\dot{e}_{kl}^{VP}\dot{e}_{kl}^{VP}} \tag{66}$$

oder die viskoplastische Arbeit

$$\dot{\eta} = \dot{W}^{VP} = \sigma_{kl}\dot{\varepsilon}_{kl}^{VP} = \sigma_V\dot{\varepsilon}_V^{VP} \tag{67}$$

ausgedrückt werden kann. Das Symbol $\langle x\rangle$ ist nach PERZYNA [14], S. 186, definiert durch

$$\langle x\rangle = \begin{cases} 0, & \text{wenn } x \leqq 0\,, \\ x & \text{wenn } x < 0\,. \end{cases} \tag{68}$$

Die Vergleichsspannung σ_V ist durch (13) festgelegt.

Die Gleichung (65) ist gegenüber [14] nur leicht in der Schreibweise modifiziert. Sie besagt, daß nach Überschreiten der Fließgrenze $\sigma_V = k_0\varkappa(\eta)$ die Fließ- (oder Kriech-)geschwindigkeit eindeutig dem „Spannungsüberschuß" zugeordnet ist. Das unterscheidet das viskoplastische Verhalten wesentlich vom zähigkeitsfreien plastischen Fließen.

Im Grenzfall $T \to 0$ geht das Gesetz (65) mit (66) gegen das zähigkeitsfreie elastisch-plastische Gesetz (18), (19) mit $k = k_0\varkappa(\varepsilon_V^p)$. Mit $\varkappa = 1$ und $H(x) = x/k_1$ folgt aus (65) die PRAGERsche dreidimensionale Verallgemeinerung ([15], S. 125) des BINGHAM-Gesetzes. Die Dehnungsverfestigungstheorie des Metallkriechens nach LUDVIK-NADAI-DAVENPORT geht ebenfalls aus (65), (66) hervor, wenn k_0 gegen Null strebt. Ein üblicher Potenzansatz (z. B. [1], S. 200) ist dann mit $H(x) = \left(\frac{x}{k_1}\right)^{\frac{1+\alpha}{\beta}}$, $\varkappa(\eta) = \left[\frac{1}{\beta}\left(\frac{3G}{k_1}\eta\right)^{\frac{1}{\beta}-1}\right]^{\frac{\beta}{1+\alpha}}$ gegeben, wobei die Normierung so gewählt wurde, daß beim einachsigen Zug unter konstanter Spannung σ_0 gerade

$$\varepsilon^{VP}(t) = \frac{\sigma_0}{3G}\left(\frac{\sigma_0}{k_1}\right)^{\alpha}\left(\frac{t}{T}\right)^{\beta} \tag{69}$$

als Lösung auftritt. Für (65) wird nur ein eindimensionaler Vektor

$$\boldsymbol{h} = \{\eta\} \tag{70}$$

benötigt, die Matrizen des Deformationsgesetzes (1), (2) im ebenen Hauptspannungszustand (10), (11) bestimmen sich zu

$$\boldsymbol{b}(\boldsymbol{\sigma}, \boldsymbol{h}) = \Phi(\boldsymbol{\sigma}, \boldsymbol{h}) \begin{Bmatrix} \sigma_1 - \frac{1}{2}\sigma_2 \\ \sigma_2 - \frac{1}{2}\sigma_1 \end{Bmatrix}, \tag{71}$$

$$\boldsymbol{c}(\boldsymbol{\sigma}, \boldsymbol{h}) = \begin{cases} \{\Phi(\boldsymbol{\sigma}, \boldsymbol{h})\,\sigma_V\} & \text{für } \eta \text{ nach (66)} \\ \{\Phi(\boldsymbol{\sigma}, \boldsymbol{h})\,\sigma_V^2\} & \text{für } \eta \text{ nach (67)} \end{cases} \tag{72}$$

mit

$$\Phi(\boldsymbol{\sigma}, \boldsymbol{h}) = \frac{2}{3}\,\frac{1+\nu}{ET}\,\frac{k_1}{\sigma_V}\left\langle H\left(\frac{\sigma_V}{\varkappa(\eta)} - k_0\right)\right\rangle. \tag{73}$$

Die Elastizitätsmatrix $\boldsymbol{J}$ wird nach (16) gebildet, die (1,2)-Matrix $\boldsymbol{A}$ ist eine Nullmatrix, für σ_V gilt wieder (14.3). Für $\varkappa$ = const. werden (70) und (72) nicht benötigt, und mit der Definition

$$h(\sigma_V) = \frac{k_1}{\sigma_V}\left\langle \mathrm{H}\left(\frac{\sigma_V}{\varkappa_0} - k_0\right)\right\rangle \tag{74}$$

ist das dann entstandene Modell des sogenannten stationären Kriechens auch als Sonderfall des viskoelastischen Gesetzes (50) enthalten, wenn $f(\sigma_V) \equiv 0$ gesetzt wird.

Das weist einerseits nochmals darauf hin, daß die Fließbedingung $\sigma_V = \varkappa_0 k_0$ dann lediglich für die Funktionsberechnung von $\boldsymbol{h}$ Bedeutung hat und keine prinzipiellen Auswirkungen auf das viskoplastische Verhalten zeigt, andererseits jedoch auch darauf, daß dieses SODERBERG-ODQVIST-Gesetz ([1], S. 188) und sein linearer Sonderfall (MAXWELL) sowie der Sonderfall eines linearen $H(x)$ mit Fließgrenze (BINGHAM-PRAGER)) nicht eindeutig den Begriffen Viskoelastizität bzw. -plastizität zuzuordnen sind. Eindeutige Unterscheidungsmerkmale treten erst mit echten Nachwirkungserscheinungen auf der Seite der Viskoelastizität bzw. mit Verfestigungserscheinungen der Viskoplastizität auf.

4.3. Material mit allgemein rheonomem Verhalten

Durch Addition der Dehnungen sind alle bisher vorgestellten Deformationsgesetze beliebig kombinierbar. Es ist leicht einzusehen, daß durch derartige Additionen der Rahmen (1), (2) nicht verlassen wird. So werden z. B. in der Arbeit [16] die Matrizen des ebenen Hauptspannungszustandes für ein Gesetz hergeleitet, welches sich aus der Addition der Dehnungen eines nichtlinear-viskoelastischen Materials nach (50) mit dem skleronom-plastischen nach (19) und einem viskoplastischen nach (65) ohne Fließgrenze ($k_0 = 0$) ergibt. Der Vektor $\boldsymbol{h}$ ist dann entsprechend $(2N + 2)$-dimensional. Die Matrizen $\boldsymbol{J}$, $\boldsymbol{A}$, $\boldsymbol{b}$ und $\boldsymbol{c}$ entstehen aus der Summation der auf die neue Dimension von $\boldsymbol{h}$ erweiterten Anteile der entsprechenden Gesetze. Die Elemente von $\boldsymbol{J}$ und einige von $\boldsymbol{A}$ bleiben bei Erreichen der Fließgrenze nur endlich, wenn das plastische Material sich verfestigt, da die Addition die Überlegungen zum plastischen Stoff nach (19) nicht beeinflußt. Dagegen existieren die inversen Gleichungen ganz analog (28), (29), so daß es sich empfiehlt, die Elemente von $\boldsymbol{C}$, $\boldsymbol{B}$, $\boldsymbol{d}$ und $\boldsymbol{e}$ analytisch zu ermitteln, falls der Grenzfall idealer Plastizität nicht ausgeschlossen werden kann. In allen anderen Fällen können diese Matrizen numerisch berechnet werden.

Für alle Beispiele war typisch, daß die Grundmatrizen nur von $\boldsymbol{\sigma}$ und $\boldsymbol{h}$, nicht aber von $\boldsymbol{\varepsilon}$ abhängen. Dieses Merkmal ist jedoch nicht zu verallgemeinern, wenn die Allgemeingültigkeit von (1), (2) und (4) nicht eingeschränkt werden soll. Darauf war bereits beim Elastizitätsgesetz, das auf der Formulierung der spezifischen elastischen Arbeit aufbaut, hingewiesen worden.

Die Dimension von $\boldsymbol{h}$ ist ein Maß dafür, welche Ordnung das gesamte Differentialgleichungssystem hat. Integraloperatoren mit beliebigen Kernen können auf Grund der Funktionaltransformation beim Übergang der diskreten Exponentialreihe auf eine kontinuierliche Koeffizientenverteilung als Differentialgesetze mit unendlichdimensonalem Vektor $\boldsymbol{h}$ gedeutet werden. Eine Überführung auf einen endlichdimensionalen Vektor $\boldsymbol{h}$ verlangt also eine diskrete Koeffizientenverteilung als genäherte Abbildung. Mit der Voraussetzung, daß diese Diskretisierung stets vorgenommen wird, ist also (1) eine Verallgemeinerung aller bisher untersuchten Deformationsgesetze.

5. Numerische Integration von Anfangs-Randwertproblemen

Es sei $\boldsymbol{\xi}$ der Vektor aller Koordinatenwerte eines Punktes des Kontinuums, $\boldsymbol{u}$ der Vektor aller Verschiebungskomponenten, gemessen von der Konfiguration des unbelasteten Körpers. Beide Vektoren sind allgemeiner aufzufassen als der Orts- und Verschiebungsvektor im geometrischen Sinne, da die Elemente von $\boldsymbol{\xi}$ auch die Parameter einer krummlinigen Darstellung sein können und in $\boldsymbol{u}$ auch Verdrehungen, wie sie z. B. in der zweidimensionalen Interpretation eines Flächentragwerkes auftreten, aufgenommen werden dürfen. Beschränken wir uns vorerst auf die geometrische lineare Theorie, so ist der Zusammenhang zwischen dem Vektor der Verschiebungsgrößen und dem der Verzerrungskomponenten durch eine lineare Beziehung

$$\boldsymbol{\varepsilon} = \boldsymbol{D}_G \boldsymbol{u} \tag{75}$$

gegeben, wobei $\boldsymbol{D}_G$ eine Matrix mit Differentialoperatoren 1. Ordnung nach den Elementen von $\boldsymbol{\xi}$ enthält, die im allgemeinen noch mit Funktionen von $\boldsymbol{\xi}$ multipliziert sind.

Die Gleichgewichtsbedingungen sollen verallgemeinert für Schnittgrößen, die in einem Vektor $\boldsymbol{q}$ zusammengefaßt sind, formuliert werden. Diese Schnittgrößen bestimmen sich linear aus den Spannungen,

$$\boldsymbol{q} = \boldsymbol{I}\boldsymbol{\sigma}\,, \tag{76}$$

wobei die Matrix $\boldsymbol{I}$ Integraloperationen mit bestimmten Integralen enthalten kann. Für Probleme, die in den Spannungen direkt formuliert sind, ist $\boldsymbol{I}$ eine Einheitsmatrix, da $\boldsymbol{q}$ und $\boldsymbol{\sigma}$ übereinstimmen. Schließlich ist die Beziehung zwischen Schnittgrößen $\boldsymbol{q}$ und eingeprägten kontinuierlich verteilten Lasten $\boldsymbol{p}$ wieder eine Differentialbeziehung 1. Ordnung,

$$\boldsymbol{D}_S \boldsymbol{q} + \boldsymbol{p} = \boldsymbol{0}\,, \tag{77}$$

wobei hier die Beschränkung auf quasistatische Vorgänge (Trägheitskräfte vernachlässigbar) erfolgt ist. Werden nun alle das Problem bestimmenden Größen zu einem Zustandsvektor $\boldsymbol{z}$ zusammengefaßt,

$$\boldsymbol{z} = \begin{Bmatrix} \boldsymbol{u} \\ \boldsymbol{\varepsilon} \\ \boldsymbol{\sigma} \\ \boldsymbol{h} \\ \boldsymbol{q} \\ \boldsymbol{p} \end{Bmatrix}, \tag{78}$$

dann ergibt sich nach der Differentiation von (75), (76), (77) nach t mit (4) folgendes System von Differentialgleichungen

$$\dot{\boldsymbol{\varepsilon}} = \boldsymbol{D}_G \dot{\boldsymbol{u}} \tag{79.1}$$

$$\dot{\boldsymbol{\sigma}} = \boldsymbol{C}(\boldsymbol{z})\,\dot{\boldsymbol{\varepsilon}} + \boldsymbol{d}(\boldsymbol{z}) \tag{79.2}$$

$$\dot{\boldsymbol{q}} = \boldsymbol{I}\dot{\boldsymbol{\sigma}} \tag{79.3}$$

$$\boldsymbol{D}_S \dot{\boldsymbol{q}} + \dot{\boldsymbol{p}} = \boldsymbol{0}\,. \tag{79.4}$$

Zusammen mit den Randbedingungen in $\boldsymbol{u}$ und $\boldsymbol{q}$ (bzw. $\dot{\boldsymbol{u}}$ und $\dot{\boldsymbol{q}}$) und den Anfangsbedingungen für $\boldsymbol{z}$ ist dieses Anfangs-Randwertproblem in folgender Weise zu lösen:

Für gegebenes $\boldsymbol{z}$ ist (79) ein lineares Randwertproblem für $\dot{\boldsymbol{z}}$, das mit zur Verfügung stehenden numerischen Methoden gelöst wird. Dabei interessiert die Vorschrift zur Bestimmung der $\dot{\boldsymbol{h}}$ nicht. Erst nach der Lösung des Randwertproblems werden diese Gleichungen ((2.) oder (6)) zusätzlich ausgewertet. Damit ist es möglich, alle Zuwüchse in ihrer Abhängigkeit von $\boldsymbol{z}$ anzugeben:

$$\dot{\boldsymbol{z}} = \boldsymbol{f}(\boldsymbol{z}, t)\,. \tag{80}$$

Das ist aber genau die Vorschrift, um mit bewährten Methoden zur numerischen Integration eines Anfangswertproblems rechnen zu können. Das Randwertproblem (79) für die Zuwüchse entspricht dem eines elastischen Körpers mit Anfangsspannungszustand und inhomogenen, orthotropen Materialeigenschaften. Alle Stoffe mit assoziiertem Fließgesetz haben dabei die gleichen Symmetrieeigenschaften der Matrix $\boldsymbol{J}$, die für elastisches Material typisch sind. Das zeigen alle Beispiele. Potentialeigenschaften der rheonomen Anteile sind nur dann von Interesse, wenn der Algorithmus (60) bis (64) eingesetzt werden muß.

Dieses Verfahren ist ohne weiteres auf geometrisch nichtlineare Aufgaben zu übertragen, die sich nur durch die Abhängigkeit von $\boldsymbol{D}_G$, $\boldsymbol{I}$ und $\boldsymbol{D}_S$ von $\boldsymbol{u}$ vom linearen Problem unterscheiden, so daß (79.1), (79.3) und (79.4) durch

$$\dot{\boldsymbol{\varepsilon}} = \bar{\boldsymbol{D}}_G(\boldsymbol{u})\,\dot{\boldsymbol{u}} \qquad = \bar{\bar{\boldsymbol{D}}}_G(\boldsymbol{z})\dot{\boldsymbol{u}} \tag{81.1}$$

$$\dot{\boldsymbol{q}} = \bar{\boldsymbol{I}}(\boldsymbol{u})\,\dot{\boldsymbol{\sigma}} + \bar{\boldsymbol{R}}_1(\boldsymbol{u}, \boldsymbol{\sigma})\,\boldsymbol{u} \qquad = \bar{\bar{\boldsymbol{I}}}(\boldsymbol{z})\,\dot{\boldsymbol{\sigma}} + \bar{\bar{\boldsymbol{R}}}_1(\boldsymbol{z})\,\dot{\boldsymbol{u}} \tag{81.2}$$

$$\bar{\boldsymbol{D}}_S(\boldsymbol{u})\,\dot{\boldsymbol{q}} + \bar{\boldsymbol{R}}_2(\boldsymbol{u}, \boldsymbol{q})\,\dot{\boldsymbol{u}} + \dot{\boldsymbol{p}} = \bar{\bar{\boldsymbol{D}}}_S(\boldsymbol{z})\,\boldsymbol{q} + \bar{\bar{\boldsymbol{R}}}_2(\boldsymbol{z})\,\dot{\boldsymbol{u}} + \dot{\boldsymbol{p}} = \boldsymbol{0} \tag{81.3}$$

ersetzt werden, ohne daß der lineare Charakter der Operatormatrizen $\bar{\bar{\boldsymbol{D}}}_G$, $\bar{\bar{\boldsymbol{I}}}$, $\bar{\bar{\boldsymbol{D}}}_S$ bezüglich $\boldsymbol{\xi}$ bei festem $\boldsymbol{z}$ verlorengeht. Der Vektor $\boldsymbol{z}$ kann stets so gewählt werden, daß die Koeffizienten vor den Operatorelementen reine Funktionen von Elementen von $\boldsymbol{z}$ darstellen. Bei geeigneter Formulierung der Gleichungen (81) kann wieder die direkte Korrespondenz der Zuwachsgleichungen zum elastischen Problem erreicht werden. So bedeutet z. B. die Verwendung von (79.1) (Unabhängigkeit der Differentialmatrix $\boldsymbol{D}$ von $\boldsymbol{z}$) den Bezug der Verzerrung auf die momentane Geometrie. Die Integration von (79.1) über t ist abhängig vom Verformungsweg und existiert nicht in geschlossener Form mit Ausnahme der körperfesten Hauptachsen, wo das Integral dem HENCKYschen Verzerrungsmaß entspricht.

Für die numerische Lösung des Randwertproblems muß der Körper in diskrete Stützstellen aufgelöst werden, so daß z dann als Zusammenfassung aller $z^{(i)} = z(\xi_i, t)$ zu verstehen ist. Die Operation $\bar{\bar{I}}$ besteht nur aus Summationsprozessen. Für die Lösung des Randwertproblems stehen folgende Verfahren zur Verfügung:

— ist das Problem eindimensional, wird es als Anfangswertproblem gelöst. Da es linear ist, entspricht diese Methode dem Übertragungsmatrizenverfahren. Beispiele für dessen erfolgreiche Anwendung auf elastisch-plastische (Deformationsgesetz nach (28)) und elastisch-viskoplastische Rotationsschalen [17], [18] sowie auf elastisch-viskoplastische Rohre und Scheiben [19] und Kreisplatten [20] und allgemein rheonome Rotationsschalen [16] liegen vor.

— nach Reduktion eines zweidimensionalen Problems läßt sich dieses Verfahren auch auf derartige Aufgaben mit einfachen Randformen übertragen. Die Reduktion kann nach Kantorowitsch [21] oder mit der Methode der Differenzen-Differentialgleichungen [22] erfolgen.

— zweidimensionale Probleme können mit dem Differenzenverfahren gelöst werden, wenn die Ränder in den gewählten Koordinatensystem Koordinatenlinien bilden. Anwendungsbeispiele liegen aber nur für geometrisch eindimensionale Aufgaben vor [23], [24].

— zwei- bzw. dreidimensionale Probleme mit komplizierten Randformen können mit der Methode der finiten Elemente gelöst werden.

Da die Lösung des Randwertproblems je nach dem Verfahren zur Integration des Anfangswertproblems ein- oder mehrfach während eines Zeitschrittes erfolgen muß, ist die Effektivität dieser Teilaufgabe von großer Bedeutung.

Für die beiden zuletzt genannten Methoden ist das Auftreten großer Gleichungssysteme mit Bandstruktur typisch. Deren Lösung erfordert beträchtliche Rechenzeiten. Daher wird zur Erhöhung der Effektivität die ständige Inversion der zustandsabhängigen Steifigkeitsmatrix umgangen, indem von einer sinnvoll gewählten konstanten Steifigkeitsmatrix ein zustandsabhängiger Rest abgespaltet wird. Dieser Rest wird dem Lastvektor zugeordnet und die Lösung iterativ ermittelt [25]. Der Gesamtaufwand wird im allgemeinen etwas geringer, da stets von der gleichen, nur einmal zu invertierenden konstanten Steifigkeitsmatrix ausgegangen wird.

Die numerische Integration des Anfangswertproblems (80) erfolgt häufig nach der sehr groben Methode des Euler-Cauchyschen Polygonzugverfahrens („step-by-step"-Verfahren). Sehr gute Erfahrungen wurden aber beim Einsatz des Runge-Kutta-Verfahrens zur Lösung von (80) gewonnen ([16] bis [22]).

Literatur

1 Работнов, Ю. Н., Ползучесть элементов конструкций, Москва 1966.

2 Bergander, H., Deformationsgesetze des Modellstruktursystems. ZAMM 52, S. 397—408 (1972).

3 Ильюшин, А. А., Победря, Б. Е., Основы математической теории термовязкоупругости, Москва 1970.

4 Качанов, Л. М., Основы теории пластичности, Москва 1969.

5 Backhaus, G., Zum Stoffgesetz des elastisch-plastischen Körpers, ZAMM 52, T 293—T 305 (1972).

6 Iwan, W. D., On a Class of Models for the Yielding Behavior of Continuous and Composite Systems, J. Appl. Mech. 34, Series E, S. 612—617 (1967).

7 Backhaus, G., Spannungs-Deformations-Beziehung für den plastischen Bereich von Metallen, Tagung Festkörpermechanik, Festigkeitslehre und Materialverhalten, Berichte, B XXXVIII, Leipzig 1976.

8 Bergander, H., Berechnungsgerechte Deformationsgesetze in der nichtlinearen Viskoelastizitätstheorie, Mitteilungen der Mathematischen Gesellschaft der DDR, H. 3—4, S. 5—38 (1975).

9 Bergander, H., Die Aufbereitung der Grundgleichungen der nichtlinearen Viskoelastizitätstheorie für die numerische Lösung von Anfangs-Randwertproblemen, Diss. B, TU Dresden, 1976.

10 Bergander, H., Nichtlineare Deformationsgesetze in der Festkörperrheologie, ZAMM 52, T 307—T 311 (1972).

11 Green, A. E.; Rivlin, R. S., The Mechanics of Non-linear Materials with Memory, Part I: Arch. Rat. Mech. Anal. 1, S. 1—21 (1957), Part II: Arch. Rat. Mech. Anal. 3, S. 82—90 (1959).

12 Bergander, H.; Balke, H.; Hartmann, D.; Lehmann, H.; Röhle, H.; Tränkner, M.; Ulbricht, V., Berechnung von Bauteilen aus Thermoplasten im nichtlinearen Bereich der Deformation, Plaste und Kautschuk 21, S. 606—611 (1974).

13 Zurmühl, R., Praktische Mathematik für Ingenieure und Physiker, Berlin, Göttingen, Heidelberg, 1961.

14 Perzyna, P., Teoria Lepkoplastycznoséi, Warschau 1966.

15 Prager, W., Einführung in die Kontinuumsmechanik, Basel, Stuttgart, 1961.

16 Röhle, H.; Ulbricht, V., Berechnung von Rotationsschalen bei nichtlinearem Deformationsgesetz, Diss., TU Dresden, 1975.

17 Herrlich, O., Die Berechnung des elastisch-plastischen Verhaltens von rotationssymmetrisch belasteten Rotationsschalen unter Berücksichtigung endlicher Verformungen, Habilitationsschrift, TU Dresden, 1969.

18 Friedrich, H., Die Berechnung drehsymmetrisch belasteter Rotationsschalen mit viskoelastischem Materialverhalten unter Berücksichtigung endlicher Verformungen, Diss., TU Dresden, 1974.

19 Wesselinow, K., Beanspruchung von dickwandigen Hohlkugeln, dünnen Kreisringscheiben und dickwandigen Hohlzylindern bei hohen Temperaturen, Diss., TU Dresden, 1974.

20 Georgi, G., Beanspruchung von dicken Kreis- und Kreisringplatten bei hohen Temperaturen, Diss., TU Dresden, 1975.

21 Kiesewetter, N., Berechnung des Spannungs- und Verformungszustandes dünner Rechteckplatten im elastisch-plastischen Materialbereich unter Berücksichtigung kleiner Verformungen, Diss., TU Dresden, 1974.

22 Weber, D., Spannungs- und Verformungsberechnung von ein- und zweifach zusammenhängenden Scheiben bei elastisch-plastischem Materialverhalten. Diss., TU Dresden, 1972.

23 Takezono, S., Inoue, K., Elasto-plastic Creep Deformations of Axisymmetrical Shells, Bull. ISME 19, S. 226—232 (1976).

24 Турусов, Р. А., Рабинович, А. Л., Нелинейная ползучесть толстостеннего полимерной трубы под действием внутреннего давления, Механика полимеров 8, стр. 493—501 (1970)

25 Koczyk, S., Grundlagen zur Berechnung der Spannungen und Verformungen bei elastisch-plastischem Materialverhalten mit der Methode der finiten Elemente, Tagung Festkörpermechanik, Festigkeitslehre und Materialverhalten, Berichte, AX, Leipzig 1976.

Eingereicht am 1. 3. 1977

Anschrift: Dr. sc. techn. Helge Bergander, Institut für Leichtbau und ökonomische Verwendung von Werkstoffen, DDR-808 Dresden, Karl-Marx-Straße, Postschließfach 44, DDR

ZAMM 58, 501–510 (1978)

D. Szeląg / Z. Mróz

Optimal Design of Elastic Beams with Unspecified Support Conditions

Es wird das Problem der optimalen Auslegung von Balken bei unbestimmter Lagerung und Steifigkeit behandelt. Dabei wird der Balken aus einzelnen Segmenten von jeweils noch freier Länge und Querschnittsfläche zusammengesetzt. Für den Fall der Auslegung nach vorgegebener Nachgiebigkeit, bei gegebener Grundfrequenz der freien Schwingungen oder Knicklast werden geeignete Optimalitätskriterien hergeleitet. Illustrative Beispiele werden vorgeführt, und der Iterationsalgorithmus für die numerische Lösung wird eingehend diskutiert.

Formulation of the problem of optimal design for beams with unspecified support position and stiffness is presented. It is assumed that the beam is composed of segments of unspecified cross-sections and lengths which are to be determined together with support positions and cross-sections. Design for specified mean compliance, first frequency of free vibrations or buckling force is considered and proper optimality criteria are derived. Illustrative examples are presented and the iterative algorithm of numerical solution is discussed in detail.

В работе рассматривается задача об оптимальном проектировании упругих балок при неопределенном расположении опор и неопределенной их жесткости. Принимается что балка состоит из сегментов с неопределенными поперечными сечениями и длинами. Эти величины следует определить совместно с расположением опор и их жесткостями. Рассматривается оптимальное проектирование при заданной глобальной жесткости системы, при заданной первой частоте свободных колебаний или заданной критической силе соотвествующей потере устойчивости. Выведены условия оптимальности и построен алгоритм итерационного решения задачи. Рассмотрены решения для нескольких частных случаев.

1. Introduction

In the present paper, we consider a more extended formulation of the optimal design problem where besides dimension design variables such as thickness variation over prescribed intervals, also location and stiffness of supports should be determined. We consider optimal design for prescribed upper level of global compliance and lower limit of the buckling load or the first frequency of free vibrations. In Sections 2, 3 and 4 the necessary optimality conditions will be derived and some simple examples will be discussed in Section 5. The iterative algorithm of numerical solution of optimization problem will be presented in Section 6.

In the previous work by Mróz and Rozvany [1], Mróz and Garstecki [2] the problem of optimal reaction of supports or optimal distribution of loading was considered using variational approach for both elastic and plastic structures. Optimal segmentation of columns was considered by Masur [4]. The present work constitutes a continuation of previous research in this directions and the variational approach developed in works [3–8] will be applied here, although we reduce the problem to determination of a set of dimension and configuration parameters.

2. Optimality conditions for global compliance design

Our analysis will be carried out for elastic beams but obvious generalizations are possible to plates or shells by using proper generalized stresses and strains. Denote by $M(x)$, $k(x)$, $w(x)$ the bending moment, curvature and deflection fields within the beam. The usual field equations are

$$M'' = -q\,, \qquad k = -w'' \tag{1}$$

where prime denotes differentiation with respect to x and q denotes the transverse loading.

Consider a beam of segmentwise constant cross-sections $A_1^b, A_2^b, \dots, A_m^b$ over segments of lengths $l_1, l_2 - l_1, \dots, l - l_{m-1}$, and supported on n supports of cross-sections $A_1^s, A_2^s, \dots, A_n^s$ at distances $s_1, s_2, \dots, s_n$ from the end of the beam, Fig. 1. Our aim is to determine optimal cross-sections of beam segments and supports, optimal division into segments and optimal support positions, such that minimize the total cost of the structure,

$$\min I = F(A_1^b, \dots, A_m^b, A_1^s, \dots, A_n^s, l_1, \dots, l_{m-1}, s_1, \dots, s_n)\,, \tag{2}$$

with the constraint imposed on global complicance of the system

$$C = \frac{1}{2}\sum_{i=1}^{m}\int\limits_{l_{i-1}}^{l_i}\frac{M^2}{EJ_i^b}\,\mathrm{d}x + \frac{1}{2}\sum_{i=1}^{n}\left[C_i^R(R_i^s)^2 + C_i^M(M_i^s)^2\right] \leqq C_0 \tag{3}$$

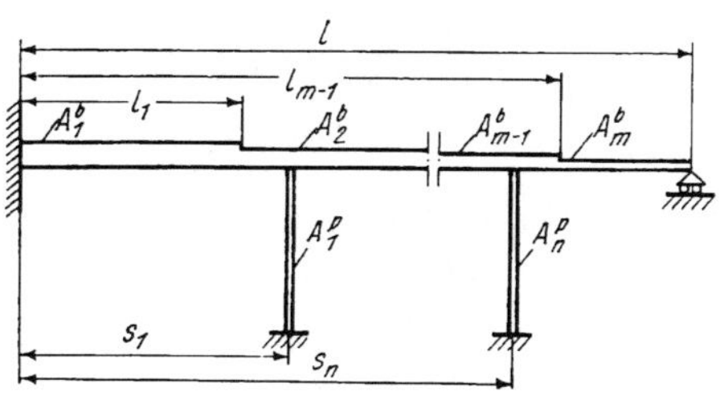

Fig. 1. Beam on elastic supports with step-wise varying stiffness

where R_i^s and M_i^s denote reactions and moments exerted by supports on the beam and C_i^R, C_i^M are support compliances which depend on support cross-sections.

In order to derive the necessary optimality conditions, let us consider the functional

$$I' = I - \mu^2(C_0 - C) \tag{4}$$

where μ^2 denotes the LAGRANGE multiplier. The variation of I' equals

$$\delta I' = \sum_{i=1}^{m} \frac{\partial F}{\partial A_i^b} \delta A_i^b + \sum_{i=1}^{m-1} \frac{\partial F}{\partial l_i} \delta l_i + \sum_{i=1}^{n} \frac{\partial F}{\partial A_i^s} \delta A_i^s + \sum_{i=1}^{n} \frac{\partial F}{\partial s_i} \delta s_i + \\ + \mu^2 \left\{ \sum_{i=1}^{m} \int_{l_{i-1}}^{l_i} \left[\frac{M\delta M}{EJ_i^b} - \frac{1}{2} \frac{M^2}{E(J_i^b)^2} \frac{\partial J_i^b}{\partial A_i^b} \delta A_i^b \right] dx + \frac{1}{2} \sum_{i=1}^{m-1} M_{l_i}^2 \left[\frac{1}{EJ_i^b} - \frac{1}{EJ_{i+1}^b} \right] \delta l_i + \right. \\ \left. + \frac{1}{2} \sum_{i=1}^{n} \left[\frac{(M_i^-)^2 - (M_i^+)^2}{EJ_{s_i}^b} \delta s_i + 2C_i^R R_i^s \delta R_i^s + 2C_i^M M_i^s \delta M_i^s + \left[(R_i^s)^2 \frac{\partial C_i^R}{\partial A_i^s} + (M_i^s)^2 \frac{\partial C_i^M}{\partial A_i^s} \right] \delta A_i^s \right] \right\}, \tag{5}$$

where M_i^- and M_i^+ denote left and right-hand values at the discontinuity and the support moment is expressed as $M_i^s = M_i^- - M_i^+$; M_{l_i} denote the values of bending moments at the segment ends $x = l_i$. Let us note that in (5) we consider the variations of both the bending moments and the design parameters. Thus in order to derive stationarity conditions, we shall use the principle of virtual work.

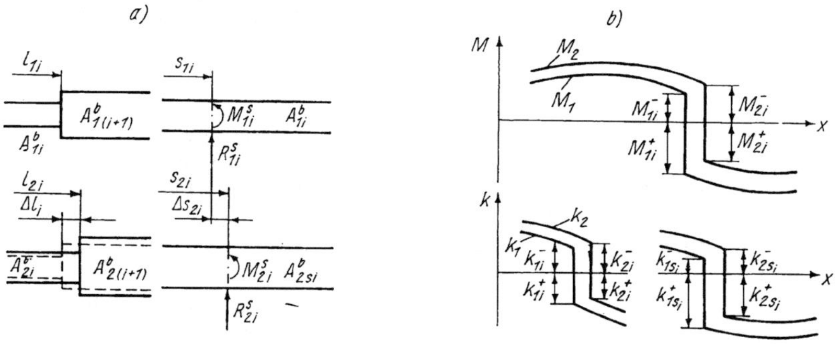

Fig. 2. a) Variation of segment and support positions, b) corresponding translation of moment and curvature discontinuities

Denoting by M_1, k_1, w_1 the static and kinematic fields for one set of design parameters and by M_2, k_2, w_2, the corresponding fields for the other set of design parameters, we can write (see Fig. 2)

$$\int_0^l M_1 k_1 \, dx = \int_0^l q_1 w_1 \, dx - \sum_{i=1}^{n} [M_{1i}^s w_1'(s_{1i}) + R_{1i}^s w_1(s_{1i})] \\ \int_0^l M_2 k_1 \, dx = \int_0^l q_2 w_1 \, dx - \sum_{i=1}^{n} [M_{2i}^s w_1'(s_{2i}) + R_{2i}^s w_1(s_{2i})] \tag{6}$$

and from (6) we obtain

$$\int_0^l (M_2 - M_1) k_1 \, dx = \int_0^l (q_2 - q_1) w_1 \, dx - \sum_{i=1}^{n} \{(M_{2i}^s - M_{1i}^s) w_1'(s_{1i}) + \\ + M_{2i}^s[w_1'(s_{2i}) - w_1'(s_{1i})] + (R_{2i}^s - R_{1i}^s) w_1(s_{1i}) + R_{2i}^s[w_1(s_{2i}) - w_1(s_{1i})]\} . \tag{7}$$

Let us note that for infinitesimal translation of supports, we can write

$$w_1(s_{2i}) = w_1(s_{1i}) + w_1'(s_{1i}) \delta s_i , \\ w_1'(s_{2i}) = w_1'(s_{1i}) + w_1''^{+}(s_{1i}) \delta s_i^+ , \tag{8}$$

where $\delta s_i = s_{2i} - s_{1i}$. Since w_1'' changes abruptly at the point of application of the concentrated moment M_i^s, we distinguish between $w_{1i}''^{+}$, δs_i^+ and $w_{1i}''^{-}$, δs_i^-. In view of (8) it is seen that

$$\int_{s_{1i}}^{s_{1i}+\delta s_i^+} (M_2 - M_1) k_1 \, dx = M_{2i}^s k_1^+ \delta s_i^+ = - M_{2i}^s w_1''^{+}(s_{1i}) \delta s_i^+ , \tag{9}$$

and similar identity for negative translation of supports δs_i^-. Thus the virtual work equation (7) can be simplified as follows:

$$\int_0^l \delta M k \, dx = \sum_{i=1}^{m} \int_{l_{i-1}}^{l_i} \frac{M\delta M}{EJ_i^b} dx = \int_0^l \delta q w \, dx - \sum_{i=1}^{n} [\delta M_i^s w_i' + \delta R_i^s w_i + R_i^s w_i' \delta s_i] \tag{10}$$

where $\delta M = M_2 - M_1$ is the variation of bending moments without including the effect of discontinuity translation.

Substituting (10) into (5) and using the relations between support reactions and their displacements and rotations

$$w_i' = C_i^M M_i^s\,, \quad w_i = C_i^R R_i^s\,, \tag{11}$$

we obtain

$$\begin{aligned}\delta I' = \sum_{i=1}^{m}\left[\frac{\partial F}{\partial A_i^b} - \frac{1}{2}\mu^2 \int\limits_{l_{i-1}}^{l_i} \frac{M^2}{E(J_i^b)^2}\frac{\partial J_i^b}{\partial A_i^b}\,\mathrm{d}x\right]\delta A_i^b + \sum_{i=1}^{m-1}\left\{\frac{\partial F}{\partial l_i} + \frac{1}{2}\mu^2 M_{l_i}^2\left[\frac{1}{EJ_i^b} - \frac{1}{EJ_{i+1}^b}\right]\right\}\delta l_i + \\ + \sum_{i=1}^{n}\left\{\frac{\partial F}{\partial A_i^s} + \frac{1}{2}\mu^2\left[(R_i^s)^2\frac{\partial C_i^R}{\partial A_i^s} + (M_i^s)^2\frac{\partial C_i^M}{\partial A_i^s}\right]\right\}\delta A_i^s + \\ + \sum_{i=1}^{n}\left\{\frac{\partial F}{\partial s_i} + \frac{1}{2}\mu^2\left[\frac{(M_i^-)^2 - (M_i^+)^2}{EJ_{s_i}^b}\right] - \mu^2 R_i^s w_i'\right\}\delta s_i\end{aligned} \tag{12}$$

and for independent variations of design variables A_i^b, l_i, A_i^s, s_i, the necessary optimality conditions are

$$\frac{\partial F}{\partial A_i^b} - \frac{1}{2}\mu^2 \int\limits_{l_{i-1}}^{l_i} \frac{M^2}{E(J_i^b)^2}\frac{\partial J_i^b}{\partial A_i^b}\,\mathrm{d}x = 0\,, \qquad i = 1, \dots, m \tag{13a}$$

$$\frac{\partial F}{\partial l_i} + \frac{1}{2}\mu^2 M_{l_i}^2\left[\frac{1}{EJ_i^b} - \frac{1}{EJ_{i+1}^b}\right] = 0\,, \qquad i = 1, \dots, m-1 \tag{13b}$$

$$\frac{\partial F}{\partial A_i^s} + \frac{1}{2}\mu^2\left[(R_i^s)^2\frac{\partial C_i^R}{\partial A_i^s} + (M_i^s)^2\frac{\partial C_i^M}{\partial A_i^s}\right] = 0\,, \qquad i = 1, \dots, n \tag{13c}$$

$$\frac{\partial F}{\partial s_i} + \frac{1}{2}\mu^2\frac{(M_i^-)^2 - (M_i^+)^2}{EJ_{s_i}^b} - \mu^2 R_i^s w_i' = 0\,, \qquad i = 1, \dots, n\,. \tag{13d}$$

In some cases it is more convenient to express these conditions in terms of kinematic quantities. Let us note that at the cross-section discontinuity $x = l_i$ the bending moment is continuous whereas curvature changes discontinuously and at the support $x = s_i$ exerting moment and force, the bending moment, shear force and curvature have discontinuities. Therefore we can write

$$M_{l_i} = k_i^- EJ_i^b = k_i^+ EJ_{i+1}^b\,, \tag{14}$$

$$M_i^- = k_{s_i}^- EJ_{s_i}^b\,, \qquad M_i^+ = k_{s_i}^+ EJ_{s_i}^b\,, \qquad R_i^s = K_i^R w_i\,, \qquad M_i^s = K_i^M w_i' \tag{15}$$

and

$$\begin{aligned}(R_i^s)^2\frac{\partial C_i^R}{\partial A_i^s} &= -\frac{(R_i^s)^2}{(K_i^R)^2}\frac{\partial K_i^R}{\partial A_i^s} = -(w_i)^2\frac{\partial K_i^R}{\partial A_i^s}\,, \\ (M_i^s)^2\frac{\partial C_i^M}{\partial A_i^s} &= -(w_i')^2\frac{\partial K_i^M}{\partial A_i^s}\end{aligned} \tag{16}$$

where k_i^-, k_i^+, $k_{s_i}^-$, $k_{s_i}^+$, M_i^-, M_i^+ are left and right-hand values at the cross-sections $x = l_i$ and $x = s_i$. The support stiffnesses are denoted by $K_i^M = 1/C_i^M$ and $K_i^R = 1/C_i^R$. In view of (14)—(16), the necessary optimality conditions take the form

$$\frac{\partial F}{\partial A_i^b} - \frac{1}{2}\mu^2 E\frac{\partial J_i^b}{\partial A_i^o}\int\limits_{l_{i-1}}^{l_i} k^2\,\mathrm{d}x = 0\,, \qquad i = 1, \dots, m \tag{17a}$$

$$\frac{\partial F}{\partial l_i} + \frac{1}{2}\mu^2 M_{l_i}[k_i^- - k_i^+] = 0\,, \qquad i = 1, \dots, m-1 \tag{17b}$$

$$\frac{\partial F}{\partial A_i^s} - \frac{1}{2}\mu^2\left[(w_i)^2\frac{\partial K_i^R}{\partial A_i^s} + (w_i')^2\frac{\partial K_i^M}{\partial A_i^s}\right] = 0\,, \qquad i = 1, \dots, n \tag{17c}$$

$$\frac{\partial F}{\partial s_i} + \frac{1}{2}\mu^2 EJ_{s_i}^b[(k_{s_i}^-)^2 - (k_{s_i}^+)^2] - \mu^2 R_i^s w_i' = 0\,, \qquad i = 1, \dots, n\,. \tag{17d}$$

Let us note that conditions (17b) for optimal segmentation of beams were first derived by MASUR [4] whereas the conditions (17a) for optimal support positions were derived by MRÓZ and ROZVANY [1].

In deriving these optimality conditions, it must be remembered that the design parameters A_i^b, l_i, A_i^s, s_i satisfy additional inequalities $A_i^b \geqq 0$, $A_i^s \geqq 0$, $0 \leqq l_i \leqq l_{i+1} \leqq l$, $0 \leqq s_i \leqq s_{i+1} \leqq l$ which are not incorporated in derivation of (12) —(17). This can be done by introducing additional LAGRANGE multipliers (see, for instance [2]). Thus it may occur that optimal support position is at the beam end or some of segments may disappear.

3. Optimal design with buckling constraint

Now let us discuss the problem of optimal design for the case of beam axially loaded by the force P and require that the critical buckling force should not be less than a prescribed value P_0. For any kinematically admissible displacement and curvature fields an upper bound of buckling force can be obtained from the RAYLEIGH quotient

$$P_{cr} \leqq \frac{\frac{1}{2} \sum_{i=1}^{m} \int_{l_{i-1}}^{l_i} EJ_i^b \bar{k}^2 \, dx + \frac{1}{2} \sum_{i=1}^{n} [K_i^R (\bar{w}_i)^2 + K_i^M (\bar{w}_i')^2]}{\frac{1}{2} \int_0^l (\bar{w}')^2 \, dx}, \tag{18}$$

where the equality sign occurs for the actual buckling mode $\bar{w} = w$, $\bar{k} = k$.

The optimal design problem can thus be formulated similarly as previously, namely, we shall opitmize the total cost (2) subject to the constraint

$$P_{cr} \geqq P_0 , \tag{19}$$

which in view of (18) can be expressed as follows:

$$-\left\{\frac{1}{2} \sum_{i=1}^{m} \int_{l_{i-1}}^{l_i} EJ_i^b k^2 \, dx + \frac{1}{2} \sum_{i=1}^{n} [K_i^R (w_i)^2 + K_i^M (w_i')^2] - \frac{1}{2} P_0 \int_0^l (w')^2 \, dx\right\} \leqq 0 . \tag{20}$$

In order to derive the necessary optimality conditions, let us consider the functional

$$I' = F - \frac{1}{2} \mu^2 \left\{\sum_{i=1}^{m} \int_{l_{i-1}}^{l_i} EJ_i^b k^2 \, dx + \sum_{i=1}^{n} [K_i^R (w_i)^2 + K_i^M (w_i')^2] - P_0 \int_0^l (w')^2 \, dx\right\} \tag{21}$$

whose variation can be expressed as follows:

$$\begin{aligned}
\delta I' = {} & \sum_{i=1}^{m} \frac{\partial F}{\partial A_i^b} \delta A_i^b + \sum_{i=1}^{m-1} \frac{\partial F}{\partial l_i} \delta l_i + \sum_{i=1}^{n} \frac{\partial F}{\partial A_i^s} \delta A_i^s + \sum_{i=1}^{n} \frac{\partial F}{\partial s_i} \delta s_i - \\
& - \mu^2 \left\{ \sum_{i=1}^{m} \int_{l_{i-1}}^{l_i} \left[EJ_i^b k \, \delta k + \frac{1}{2} E k^2 \frac{\partial J_i^b}{\partial A_i^b} \delta A_i^b \right] dx + \frac{1}{2} \sum_{i=1}^{m-1} [EJ_i^b (k_i^-)^2 - EJ_{i+1}^b (k_i^+)^2] \, \delta l_i + \right. \\
& + \frac{1}{2} \sum_{i=1}^{n} \left[EJ_{s_i}^b [(k_{s_i}^-)^2 - (k_{s_i}^+)^2] \, \delta s_i + \left[\frac{\partial K_i^R}{\partial A_i^s} (w_i)^2 + \frac{\partial K_i^M}{\partial A_i^s} (w_i')^2 \right] \delta A_i^s \right] + \\
& \left. + \sum_{i=1}^{n} [K_i^R w_i \, \delta w_i + K_i^M w_i' \, \delta w_i'] - P_0 \int_0^l w' \, \delta w' \, dx \right\} .
\end{aligned} \tag{22}$$

In deriving (22), we have used the principle of virtual work

$$\begin{aligned}
& \sum_{i=1}^{m} \int_{l_{i-1}}^{l_i} EJ_i^b k \, \delta k \, dx + \sum_{i=1}^{m-1} M_{l_i} [k_i^- - k_i^+] \, \delta l_i + \sum_{i=1}^{n} EJ_{s_i}^b [(k_{s_i}^-)^2 - (k_{s_i}^+)^2] \, \delta s_i \\
& = \int_0^l q \, \delta w \, dx - \sum_{i=1}^{n} [R_i^s \, \delta w_i - R_i^s w_i' \, \delta s_i + M_i^s \, \delta w_i']
\end{aligned} \tag{23}$$

where $\delta k = k_2 - k_1$ is the variation of curvature field without including effect of translation of concentrated bending moments M_i^s at support positions. Moreover considering the term $(- P_0 w'')$ as fictitious lateral loading $q = -P_0 w''$, we have

$$\int_0^l q \, \delta w \, dx = - P_0 \int_0^l w'' \, \delta w \, dx = - P_0 [w' \, \delta w]_0^l + P_0 \int_0^l w' \, \delta w' \, dx \tag{24}$$

and it will be assumed that $[w' \, \delta w]_0^l = 0$. Using (24) and (23), from (22) we obtain the following optimality conditions

$$\frac{\partial F}{\partial A_i^b} - \frac{1}{2} \mu^2 \frac{\partial J_i^b}{\partial A_i^b} \int_{l_{i-1}}^{l_i} E k^2 \, dx = 0 , \qquad i = 1, \dots, m \tag{25a}$$

$$\frac{\partial F}{\partial l_i} + \frac{1}{2} \mu^2 M_{l_i} [k_i^- - k_i^+] = 0 , \qquad i = 1, \dots, m-1 \tag{25b}$$

$$\frac{\partial F}{\partial A_i^s} - \frac{1}{2} \mu^2 \left[(w_i)^2 \frac{\partial K_i^R}{\partial A_i^s} + (w_i')^2 \frac{\partial K_i^M}{\partial A_i^s} \right] = 0 , \qquad i = 1, \dots, n \tag{25c}$$

$$\frac{\partial F}{\partial s_i} + \frac{1}{2} \mu^2 EJ_{s_i}^b [(k_{s_i}^-)^2 - (k_{s_i}^+)^2] - \mu^2 R_i^s w_i' = 0 , \qquad i = 1, \dots, n . \tag{25d}$$

It is seen that the stationarity conditions (25) are identical to conditions (17) and they can also be presented in the form (13).

4. Optimal design with specified first frequency of free vibrations

For harmonic vibrations, one of design constraints can be set by requiring that the first eigenfrequency be not less than prescribed value [2–4]. For any kinemttically admissible fields $\overline{w}$ and $\overline{k}$ the upper bound of the first frequency is defined by the formula

$$\omega^2 \leqq \frac{\frac{1}{2}\sum\limits_{i=1}^{m}\int\limits_{l_{i-1}}^{l_i} EJ_i^b\overline{k}^2\,\mathrm{d}x + \frac{1}{2}\sum\limits_{i=1}^{n}[K_i^R(\overline{w}_i)^2 + K_i^M(\overline{w}_i')^2]}{\frac{1}{2}\sum\limits_{i=1}^{m}\int\limits_{l_{i-1}}^{l_i} m_i\overline{w}^2\,\mathrm{d}x} \tag{26}$$

where $\overline{w}(x)$ and $\overline{k}(x)$ can be regarded as kinematically admissible amplitudes of free vibrations and m_i denotes the specific mass per unit length of beam. The equality in (26) occurs for the actual vibration mode $\overline{w} = w$, $\overline{k} = k$.

We shall look for minimum of the cost function (2) subject to the constraint

$$\omega^2 \geqq \omega_0^2 \tag{27}$$

which in view of (26) can be expressed as follows:

$$-\left\{\frac{1}{2}\sum_{i=1}^{m}\int\limits_{l_{i-1}}^{l_i} EJ_i^bk^2\,\mathrm{d}x + \frac{1}{2}\sum_{i=1}^{n}[K_i^R(w_i)^2 + K_i^M(w_i')^2] - \frac{1}{2}\omega_0^2\sum_{i=1}^{m}\int\limits_{l_{i-1}}^{l_i} m_iw^2\,\mathrm{d}x\right\} \leqq 0\,. \tag{28}$$

Similarly as previously, we shall consider the functional

$$I' = F - \mu^2\left\{\frac{1}{2}\sum_{i=1}^{m}\int\limits_{l_{i-1}}^{l_i} EJ_i^bk^2\,\mathrm{d}x + \frac{1}{2}\sum_{i=1}^{n}[K_i^R(w_i)^2 + K_i^M(w_i')^2] - \frac{1}{2}\omega_0^2\sum_{i=1}^{m}\int\limits_{l_{i-1}}^{l_i} m_iw^2\,\mathrm{d}x\right\} \tag{29}$$

where μ^2 is the LAGRANGE multiplier. Variation of (29) equals

$$\begin{aligned}
\delta I' = {} & \sum_{i=1}^{m}\frac{\partial F}{\partial A_i^b}\,\delta A_i^b + \sum_{i=1}^{m-1}\frac{\partial F}{\partial l_i}\,\delta l_i + \sum_{i=1}^{n}\frac{\partial F}{\partial A_i^s}\,\delta A_i^s + \sum_{i=1}^{n}\frac{\partial F}{\partial s_i}\,\delta s_i + \\
& + \mu^2\left\{\sum_{i=1}^{m}\int\limits_{l_{i-1}}^{l_i}\left[EJ_i^bk\,\delta k + \frac{1}{2}Ek^2\frac{\partial J_i^b}{\partial A_i^b}\,\delta A_i^b\right]\mathrm{d}x + \frac{1}{2}\sum_{i=1}^{m-1}[EJ_i^b(k_i^-)^2 - EJ_{i+1}^b(k_i^+)^2]\,\delta l_i + \right. \\
& + \sum_{i=1}^{n}\left[\frac{1}{2}EJ_{s_i}^b[(k_{s_i}^-)^2 - (k_{s_i}^+)^2]\,\delta s_i + K_i^Rw_i\,\delta w_i + K_i^Mw_i'\,\delta w_i' + \right. \\
& + \left.\frac{1}{2}\left[(w_i)^2\frac{\partial K_i^R}{\partial A_i^s} + (w_i')^2\frac{\partial K_i^M}{\partial A_i^s}\right]\delta A_i^s\right] - \omega_0^2\sum_{i=1}^{m}\int\limits_{l_{i-1}}^{l_i}\left[m_iw\,\delta w + \frac{1}{2}w^2\frac{\partial m_i}{\partial A_i^b}\,\delta A_i^b\right]\mathrm{d}x - \\
& \left. - \sum_{i=1}^{m-1}\omega_0^2w_{l_i}[m_i - m_{i+1}]\,\delta l_i\right\}.
\end{aligned} \tag{30}$$

Now, if the fictitious lateral loading equals $q = m\omega_0^2w^2$, the virtual work equation takes the form

$$\begin{aligned}
& \sum_{i=1}^{m}\int\limits_{l_{i-1}}^{l_i} EJ_i^bk\,\delta k + \sum_{i=1}^{m-1}M_{l_i}[k_i^- - k_i^+]\,\delta l_i + \sum_{i=1}^{n}EJ_i^b\,[(k_{s_i}^-)^2 - (k_{s_i}^+)^2]\,\delta s_i = \\
& = \omega_0^2\sum_{i=1}^{m}\int\limits_{l_{i-1}}^{l_i} m_iw\,\delta w\,\mathrm{d}x - \sum_{i=1}^{n}[R_i^s\,\delta w_i + M_i^s\,\delta w_i' - R_i^sw_i'\,\delta s_i]
\end{aligned} \tag{31}$$

where as previously, the variation $\delta k(x)$ does not include the translation of concentrated couple at supports.

Substituting (31) into (30), we obtain the following optimality conditions

$$\frac{\partial F}{\partial A_i^b} - \frac{1}{2}\mu^2\int\limits_{l_{i-1}}^{l_i}\left[Ek^2\frac{\partial J_i^b}{\partial A_i^b} - \omega_0^2w^2\frac{\partial m_i}{\partial A_i^b}\right]\mathrm{d}x = 0\,, \qquad i = 1, \ldots, m \tag{32a}$$

$$\frac{\partial F}{\partial l_i} + \frac{1}{2}\mu^2\,\{M_{l_i}[k_i^- - k_i^+] + \omega_0^2w_{l_i}^2[m_i - m_{i+1}]\} = 0\,, \qquad i = 1, \ldots, m-1 \tag{32b}$$

$$\frac{\partial F}{\partial A_i^s} - \frac{1}{2}\mu^2\left[(w_i)^2\frac{\partial K_i^R}{\partial A_i^s} + (w_i')^2\frac{\partial K_i^M}{\partial A_i^s}\right] = 0\,, \qquad i = 1, \ldots, n \tag{32c}$$

$$\frac{\partial F}{\partial s_i} + \frac{1}{2}\mu^2EJ_{s_i}^b\,[(k_{s_i}^-)^2 - (k_{s_i}^+)^2] - \mu^2R_i^sw_i' = 0\,, \qquad i = 1, \ldots, n\,. \tag{32d}$$

The stationarity conditions (32) remain valid also in the case of existence of discrete masses on the beam. Applicability of these conditions in simple cases will be illustrated in the next section.

5. Examples

In the following, we shall consider the beam of constant cross-section and discuss only optimal design of supports. Let the total cost be linear function of cross-section areas of the beam and the support

$$I = c_b' A^b l + c_s' A^s H = c_b A^b + c_s A^s \,. \tag{33}$$

We shall look for optimal position of the support and optimal stiffness of the beam and support $K^b = EJ^b$ and $K^s = EA^s/H$, for different cost ratios c_b/c_s. It will be assumed that $J^b = cA^b$ which is valid for the beam of constant height, although any non-linear relationship could also be used.

5.1. Beams designed for specified mean compliance

Let us consider a cantilever beam which is to be supported within its span $0 \leqq x \leqq l$. Assume the support to have rectangular cross-section of height h. For the uniformly loaded beam, the optimality conditions (17) yield

$$c_b = \frac{c\mu^2}{2E(J^b)^2} \int\limits_0^l M^2 \,\mathrm{d}x \,, \qquad c_s = \mu^2[(R^s)^2/2 + 2(M^s)^2/h^2]/EA^s \,, \tag{34a), (34b}$$

$$\{[M^-(s) + M^+(s)]\, J^s/2J^bH - R\}\, w'(s) = 0 \,. \tag{34c}$$

It can be shown that the optimal support condition $w'(s) = 0$, $M^s = 0$ is satisfied and the support exerts only reaction R^s. The optimal solution is presented in Fig. 3 where dimensionless variables $C_k = c_b l^3/cc_s H$, $Z = J^b H/A^s l^3$, $\bar{s} = s/l$ are used. It is seen that for $0 \leqq C_k \leqq 19.2$ the optimal position of support is at the end of beam and $Z = 0.0417$ corresponds to the optimal design. When the beam cost factor c_b increases, its relative stiffness Z decreases and the support moves to the asymptotic position $\bar{s} = 0.710$. For $C_k < 3.2$ the end support disappears and the optimal design with free end is obtained.

The optimal solution for the beam simply supported at one end and clamped at other end with internal support to be determined is presented in Fig. 4. It is seen that for $0 \leqq C_k \leqq 106$ the internal support vanishes and its optimal position for $C_k > 106$ varies only within a vary narrow interval.

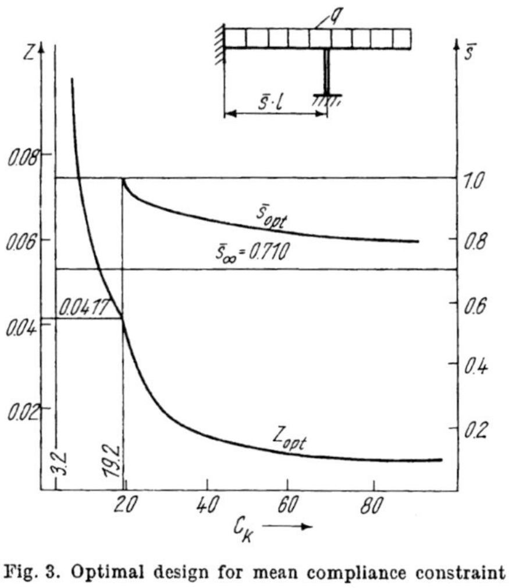

Fig. 3. Optimal design for mean compliance constraint

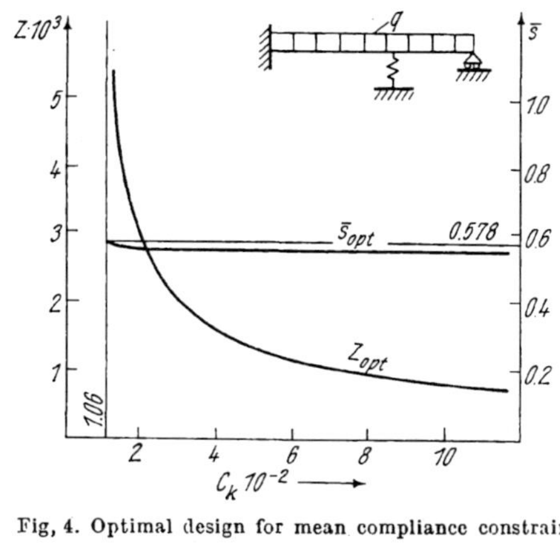

Fig. 4. Optimal design for mean compliance constraint

5.2 Optimal design of beam with support for specified buckling force

The necessary optimality conditions for the beam loaded only by the axial force P and for the cost function (33), have the form

$$c_b = \frac{\mu^2}{2c} \int\limits_0^l E(w'')^2 \,\mathrm{d}x \,, \qquad c_s = \mu^2 E \frac{w_s^2}{2H} \,, \qquad w_s' R^s = 0 \tag{35}$$

where w_s denotes the deflection at the support corresponding to the first buckling mode.

Figure 5 presents the variation of the nondimensional buckling force $P = Pl^2/EJ^b\pi^2$ with $\bar{s} = s/l$ for different ratios $Z = J^bH/A^sl^3$ for the beam with simply supported and clamped ends. The broken line connects the optimal solutions for which $w_s'R^s = 0$. It turns out that for $Z < Z^l = 4.281 \cdot 10^{-3}$ the optimal support position corresponds to the condition $w_s' = 0$ whereas for $Z > Z^l$ there is $R^s = 0$ at the optimal position and support is located at the node of the second eigenmode. For $Z = Z^l$ the two modes correspond to the same buckling load and these modes are shown in Fig. 6. The variation of the beam inertia moment $\bar{J} = J^b/J_0$ and the support cross-section area $\bar{A} = A^sl^3/HJ_0\pi^2$ for a given buckling load $P_{cr} = \text{const}$ are shown in Fig. 7a; here EJ_0 denotes the minimal beam stiffness for prescribed buckling load. The dotted line represents the constant cost of structure. Fig. 7b shows the evolution of optimal design for varying ratios C_k. It is seen that for $C_k' \leqq C_k \leqq C_k''$ the internal support of varying stiffness exists whereas for $C_k \leqq C_k'$ this support vanishes and for $C_k \geqq C_k''$ the support stiffness equals $Z = Z^l$ and then $R^s = 0$. Fig. 7c shows the optimal design for similar cross-sections $(J^b = c(A^b)^2)$ in the plane $(\overline{A^b} = A^b/A_0, \bar{A})$. The dotted lines represent the constant cost of the structure. It is seen that because of concavity of the line AB representing designs with $w_s' = 0$, the optimal solution corresponds either to A or B; thus either support or its reaction vanishes and there are no solutions corresponding to $w_s' = 0$.

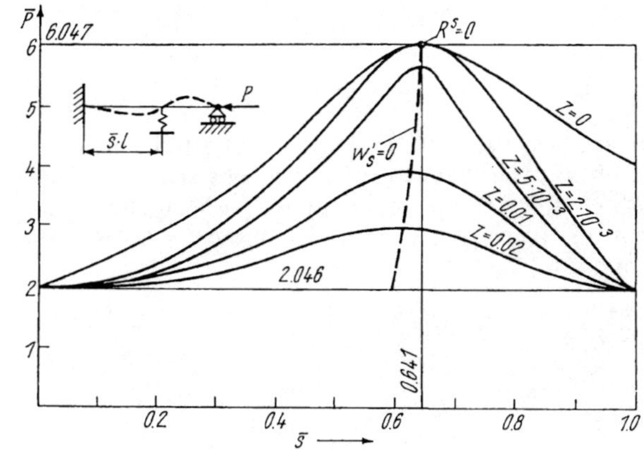

Fig. 5. Optimal design for maximum buckling force: variation of $\bar{P} = Pl^2/EJ^b\pi^2$ with support position

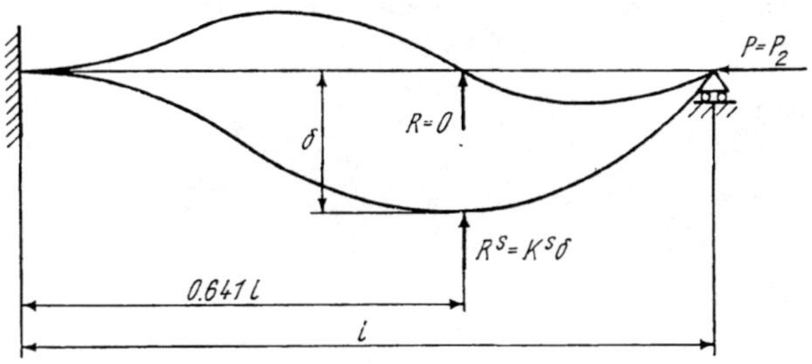

Fig. 6. Two buckling modes for the same value of critical force

a)

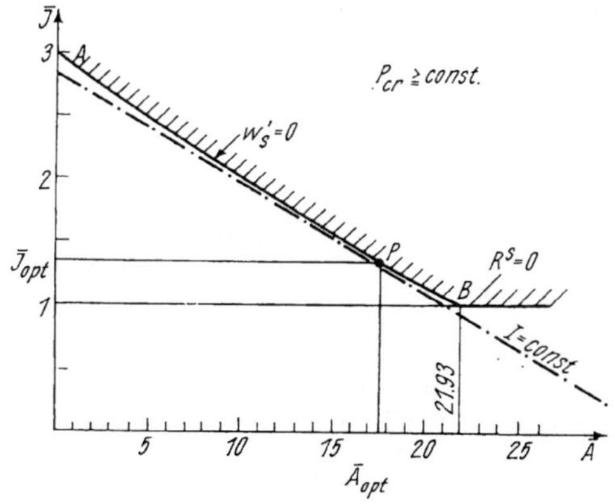

b)

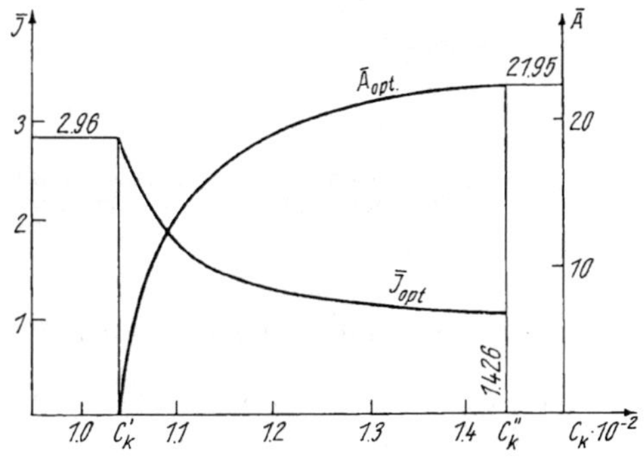

c)

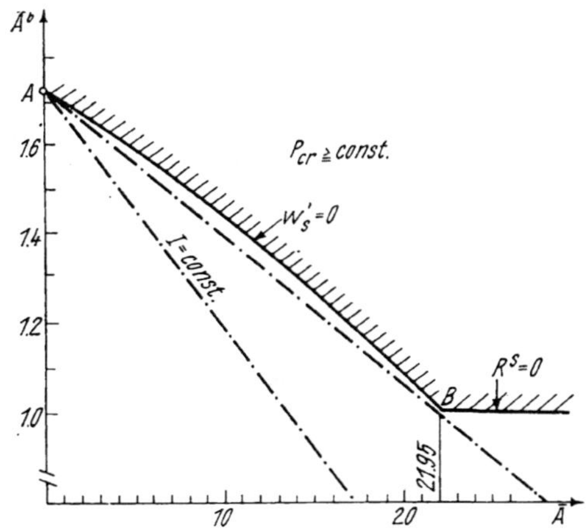

Fig. 7. a) Variation of support cross-section area $\bar{A} = A^s l^3/HJ_0\pi^2$ and the beam inertia moment $J = J^b/J_0$ for constant critical force and for $J^b = cA^b$, b) dependence of $\bar{A}_{opt}$ and $\bar{J}_{opt}$ on cost ratio $C_k = c_b l^3/cc_s H$, c) variation of $\overline{A^b} = A^b/A_0$ and $\bar{A}$ for constant critical force and for $J^b = c(A^b)^2$

5.3 Optimal design of beams for given first frequency of free vibrations

The optimality conditions for the total cost (33) have the form

$$c_b = \left(\frac{\mu^2}{2}\right) \int\limits_0^l \left[\frac{cM^2}{E(J^b)^2} - \omega_0^2 \varrho w^2\right] \mathrm{d}x\,, \tag{36a}$$

$$c_s = \mu^2 E \frac{w_s^2}{2H}, \qquad w_s' R^s = 0 \tag{36b}$$

where $\varrho = m/A^b$. Solving the characteristic equation for the beam with simply supported and clamped ends for different values of stiffness of beam and support, we obtain the diagram of Fig. 8, where $\beta = (\varrho A^b \omega_0^2 l^4/EJ^b)^{1/4}$, $Z = J^b H/A^s l^3$. Similarly as previously, for $Z \leqq Z^l = 7.26 \cdot 10^{-4}$ the optimal support position corresponds to $w_s' = 0$, whereas for $Z > Z^l$ the support is located at the node of the second eigenmode and $R^s = 0$. For $Z = Z^l$, the two eigenmodes correspond to the same value of free frequency.

Using the virtual work equation, we have

$$\int\limits_0^l \frac{M^2}{EJ^b} \mathrm{d}x - \int\limits_0^l \omega_0^2 \varrho A^b w^2 \,\mathrm{d}x = 0 \tag{37}$$

and it is seen that the right-hand term of (36a) vanishes independently on value of the beam cross-section area. Thus there is no optimal solution in the class of beams for which $J^b = cA^b$ since variations of the cross-section area and support stiffness can reduce the total cost without affecting the free frequency.

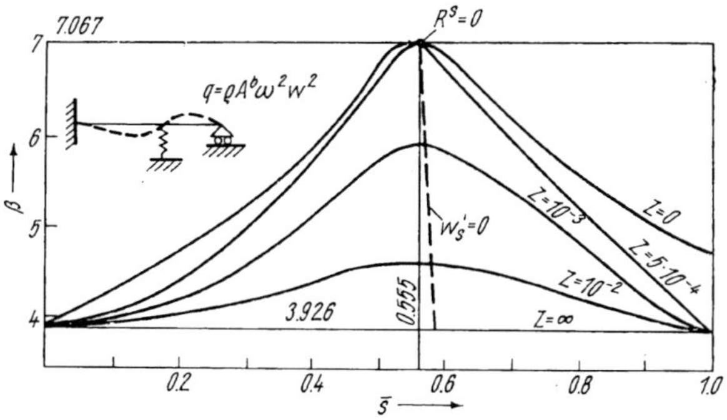

Fig. 8. Optimal design for maximum fundamental frequency of free vibrations: variation of $\beta = (\varrho A^b \omega_0^2 l^4/EJ^b)^{1/4}$ with support position $\bar{s} = s/l$

6. Iterative generation of optimal design

The optimality conditions (13), (17), (25) or (32) constitute a set of non-linear transcendent equations whose solutions in more complex cases is possible through iterative procedure. In this section, we shall discuss one of such procedures which could be applied in generating optimal solution for selected set of design variables and constraints on mechanical behaviour.

Consider a beam for which the relation between the moment of inertia and the cross-section area is a homogeneous function of order r:

$$J_i^b = c(A_i^b)^r\,, \qquad i = 1, \ldots, m\,. \tag{38}$$

Let the total cost be a sum of material volume of the beam and the supports:

$$F = V = V^b + \sum_{i=1}^n V_i^s = \sum_{i=1}^m A_i^b(l_i - l_{i-1}) + \sum_{i=1}^n H_i A_i^s\,. \tag{39}$$

The optimality conditions (32) for optimal design with frequency constraint have the form

$$\lambda^2 = \left[r \cdot \int\limits_{l_{i-1}}^{l_i} \frac{M^2}{EJ_i^b} \mathrm{d}x - \int\limits_{l_{i-1}}^{l_i} \omega^2 \varrho A_i^b w^2 \,\mathrm{d}x\right] [A_i^b(l_i - l_{i-1})]^{-1}\,, \qquad i = 1, \ldots, m \tag{40a}$$

$$\lambda^2 = \frac{M_{l_i}^2}{EJ_i^b J_{i+1}^b} (J_{i+1}^b - J_i^b)\,(A_{i+1}^b - A_i^b)^{-1} - \omega^2 \varrho w_{l_i}^2\,, \qquad i = 1, \ldots, m-1 \tag{40b}$$

$$\frac{1}{EJ^b_{si}}[(M_i^-)^2 - (M_i^+)^2] - 2R_i^s w_i' = 0\,, \qquad i = 1, \dots, n \tag{40c}$$

$$\lambda^2 = -\frac{1}{H_i}\left[(R_i^s)^2 \frac{\partial C_i^R}{\partial A_i^s} + (M_i^s)^2 \frac{\partial C_i^M}{\partial A_i^s}\right], \qquad i = 1, \dots, n \tag{40d}$$

where $m_i = \varrho A_i^b$, $\lambda^2 = 2/\mu^2$. When $\omega = 0$, equations (40) become the optimality conditions for the case of mean compliance or stability constraint. Let us note that the same optimality conditions would be obtained for a dual problem when for a structure of constant volume the free frequency or buckling force is maximized. It is more convenient to carry out numerical computations for the dual problem and we shall discuss the numerical procedure for this problem.

Since different optimality conditions apply to design variables A_i^b, l_i, A_i^s, s_i, the iterative procedure will be carried out sequentially for each design variable, regarding the remaining variables as parameters.

Let us start with considering determination of segmentation variables l_i with other variables fixed. The iterative method of solution of the optimality conditions (40b) was proposed by MASUR [4]. The consecutive approximation $l_{i(p+1)}$ is obtained by using the static field from the previous iterative step. A similar approach can be used for the dual formulation.

Assuming $l_i = l_{i(p)}$ to be known, let us calculate the right-hand side of (40b):

$$\lambda^2_{i(p)} = \left.\frac{M^2_{l_i}}{EJ_i^b J_{i+1}^b}\,\frac{J_{i+1}^b - J_i^b}{A_{i+1}^b - A_i^b} - \omega^2 \varrho w^2_{l_i}\right|_{l_i = l_{i(p)}}. \tag{41}$$

New segment position will be obtained by requiring that the conditions (40b) be satisfied for the moment and deflection fields of the previous iteration, continued at the segment interfaces, thus

$$\begin{aligned} &\widehat{M}_{l_i} = M_{l_i} + M'_{l_i}\,\Delta l_i\,, \qquad \widehat{w}_{l_i} = w_{l_i} + w'_{l_i}\,\Delta l_i\,,\\ &\Delta l_i = l_{i(p+1)} - l_{i(p)} \end{aligned} \tag{42}$$

and

$$\lambda^2_{(p)} = \frac{\widehat{M}^2_{l_i}}{EJ_i^b J_{i+1}^b}\,\frac{J_{i+1}^b - J_i^b}{A_{i+1}^b - A_i^b} - \omega^2 \varrho \widehat{w}^2_{l_i}\,. \tag{43}$$

Neglecting higher-order terms of Δl_i from (41), (42) and (43), we obtain the following expression for the increments Δl_i

$$\Delta l_i = \frac{\lambda^2_{(p)} - \lambda^2_{i(p)}}{2\left[\dfrac{M_{l_i}M'_{l_i}}{EJ_i^b J_{i+1}^b}\,\dfrac{J_{i+1}^b - J_i^b}{A_{i+1}^b - A_i^b} - \omega^2 \varrho w_{l_i} w'_{l_i}\right]} \tag{44}$$

and the value of $\lambda^2_{(p)}$ can be determined from the condition of constant volume

$$\Delta V^b = \sum_{i=1}^{m-1} (A_{i+1}^b - A_i^b)\,\Delta l_i = 0\,. \tag{45}$$

Hence, we have

$$\lambda^2_{(p)} = \frac{\sum\limits_{i=1}^{m-1} \lambda^2_{i(p)}[\dots]\,(A_{i+1}^b - A_i^b)}{\sum\limits_{i=1}^{m-1} [\dots]\,(A_{i+1}^b - A_i^b)}\,. \tag{46}$$

Upon determining $\lambda^2_{(p)}$ from (46), it is easy to determine new segment positions from (44). Thus the iterative process is proposed and optimal beam segmentation can be determined.

The next suboptimization procedure should be elaborated in order to determine the optimal cross-section areas A_i^b for fixed values of other design variables. For the p-th approximation $A^b_{i(p)}$, the right-hand side of (40a) can be calculated,

$$\lambda^2_{i(p)} = \left\{\int\limits_{l_{i-1}}^{l_i} \left[\frac{rM^2}{EJ^b_{i(p)}} - \omega^2 \varrho A^b_{i(p)} w^2\right] \mathrm{d}x\right\} [A^b_{i(p)}(l_i - l_{i-1})]^{-1} \tag{47}$$

and the new values of $A^b_{i(p+1)}$ will be obtained by requiring that

$$\lambda^2_{(p)} = \left[\int\limits_{l_{i-1}}^{l_i} \left(\frac{rM^2}{EJ^b_{i(p)}} - \omega^2 \varrho A^b_{i(p)} w^2\right) \mathrm{d}x\right] [A^b_{i(p+1)}(l_i - l_{i-1})]^{-1}\,, \tag{48}$$

hence

$$\Delta A_i^b = A^b_{i(p+1)} - A^b_{i(p)} = A^b_{i(p)}\left[\frac{\lambda^2_{i(p)}}{\lambda^2_{(p)}} - 1\right] \tag{49}$$

and from the condition of constant volume it follows that

$$\lambda^2_{(p)} = \frac{\sum_{i=1}^{m} \lambda^2_{i(p)} A^b_{i(p)}(l_i - l_{i-1})}{\sum_{i=1}^{m} A^b_{i(p)}(l_i - l_{i-1})} . \tag{50}$$

Thus from (49) and (50) the new values $A^b_{i(p+1)}$ can be determined. The analogous procedure can be applied in determining optimal stiffnesses of supports.

Finally, let us discuss the iterative procedure that would provide optimal support positions. Assuming that supports exert only reactions R^s_i, the optimality conditions take the form

$$w'_i R^s_i = 0 , \qquad i = 1, \dots, n . \tag{51}$$

Thus either $R^s_i = 0$ or $w'_i = 0$. In the case of stability or free vibration problems it was shown that for supports of stiffness less than a critical value there is $w'_i = 0$, $R^s_i \neq 0$ whereas for stiffnesses larger than the critical value the condition $R^s_i = 0$, $w'_i \neq 0$ applies and the support is located at the node of a higher eigenmode.

The solution of the equation $w'(s_i) = 0$ is obtained by using the proper series expansion

$$w'(s_{i(p+1)}) = w'(s_{i(p)}) + w''(s_{i(p)}) \, \Delta s_i \tag{52}$$

and the increment Δs_i is calculated from the condition $w'(s_{i(p+1)}) = 0$, thus

$$\Delta s_i = s_{i(p+1)} - s_{i(p)} = \left. \frac{w'_i}{w''_i} \right|_{s_i = s_{i(p)}} \tag{53}$$

where $s_{i(p)}$ denotes the support position for p-th iteration. Assume now that we have solved particular sub-problems. Each solution would correspond to a different value of the multiplier λ^2. Let the solutions of (40a), (40b) (40d) correspond to the values $\lambda^2_b, \lambda^2_l, \lambda^2_s$ whereas for the optimal solution we require that $\lambda^2_b = \lambda^2_l = \lambda^2_s = \lambda^2$. The coordination variable λ^2 can be determined by proper rescaling design variables A^b_i, l_i, A^s_i subject to the condition of constant total cost.

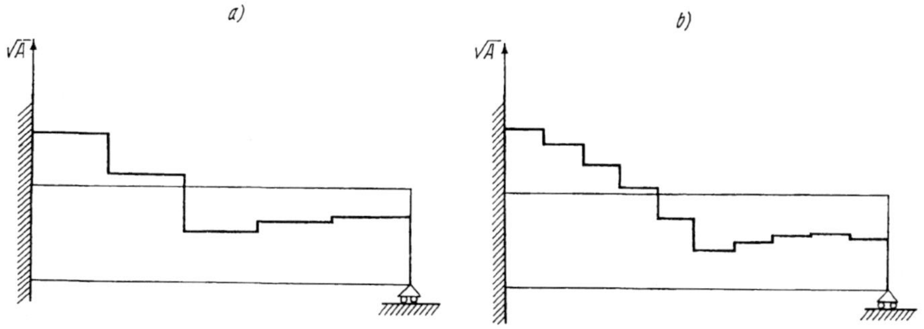

Fig. 9. Iteratively generated optima design of vibrating beam with similar cross-sections a) division into 5 segments, b) division into 10 segments

Figure 9 shows the iteratively generated optimal designs for a beam with similar cross-sections ($r = 2$) composed of segments of equal lengths. The procedure started from a uniform beam and consecutive cross-section areas were determined from the formula $A^b_{i(p+1)} = A^b_{i(p)} + \eta \, \Delta A^b_i$ where ΔA^b_i is defined by (49). The value of the coefficient η varies from $\eta = 0.3$ for first iterations to $\eta = 0.1$ for last iterations. It was found that convergence of the iterative procedure is sensitive to the value of η. With proper selection of η, the optimal solution is generated after 8—12 iterations. For a fivesegment beam (Fig. 9a) the ratio of free frequency ω_5 and the frequency ω_0 for a uniform beam equals $\omega_5/\omega_0 = 1.32$ whereas for a tensegment beam (Fig. 9b) this ratio equals $\omega_{10}/\omega_0 = 1.40$.

Acknowledgement: This research project was supported by grant No INT 75-08722 made available from Maria Skłodowska Curie fund established by contributions of the United States and Polish Governments.

References

1 Mróz, Z., Rozvany, G. I. N., Optimal design of structures with variable support conditions, J. Opt. Theory and Applications, vol. 15 (1975) pp. 85—101.
2 Mróz, Z., Garstecki, A., Optimal design of structures with unspecified loading distribution, J. Opt. Theory and Applications, vol. 20 (1976) pp. 359—380.
3 Prager, W., Taylor, J. E., Problems of optimal structural design, J. Appl. Mech., vol. 35 (1968) pp. 102—106.
4 Masur, E. F., Optimal placement of available sections in structural eigenvalue problems, J. Opt. Theory and Appl., vol. 15 (1975) pp. 69—84.
5 Karihaloo, B. L., Niordson, F. I., Optimal design of vibrating cantilevers, J. Opt. Theory and Appl., vol. 11 (1973) pp. 638—654.
6 Olhoff, N., Optimization of vibrating beams with respect to higher-order natural frequencies, DCAMM Rep. No. 83 (1975).
7 Mróz, Z., Multiparameter optimal design of plates and shells, J. Struct. Mech., vol. 1 (1973) pp. 371—392.
8 Rozvany, G. I. N., Mróz, Z., Column design: optimization of support conditions and segmentation., J. Struct. Mech. (1977).

Eingereicht am 31. 1. 1977

Address: D. Szeląg, Z. Mróz, Polish Academy of Sciences, Institute of Fundamental Technological Research, Świetokrzyska 21, 00-049 Warsaw, Poland

ZAMM 58, 511 – 518 (1978)

V. A. SAVA

The Initial-Boundary-Value Problems in the Theory of Micropolar Fluids

Der Autor betrachtet die Existenz von Lösungen eines Anfangs-Randwert-Problems für die Gleichungen, welche die Strömung inkompressibler mikropolarer Flüssigkeiten beschreiben, indem er ein vereinfachtes System von Gleichungen benutzt und dann zur Grenze übergeht.

The author considers the existence of solutions of the initial-boundary-value problem for the equations which describe the flow of incompressible micropolar fluids, by using a related, simpler set of equations and then taking a limit.

Автор исследует существание решений задачи с начальными и граничными значениями для уравнений, описывающих потоки микрополярных несжимаемых жидкостей, используя семейство более простых уравнений, связанных с предыдущими, — после чего имеет место предельный переход.

1. Introduction

Micropolar fluids, as introduced by ERINGEN [1], have a capacity for sustaining stress and body moments and have a microstructure associated with spin inertia so that microstructure and macroscopic behaviour are coupled. FÖRSTE [2], [3] has studied the existence of solutions of the boundary-value problem and of the initial-boundary-value problem for the equations which describe the flow of incompressible micropolar fluids. SAVA [4] has derived existence and uniqueness theorems of the boundary-value problem for the equations which describe the flow of incompressible micropolar fluids. Existence theorems are derived by the method of potentials in [5]. LANGE [6] has considered the existence of solutions by functional analytic methods.

This paper is devoted to the proof of existence of solution of the initial-boundary value problem for the equations which describe the flow of incompressible micropolar fluids, by using a related, simpler set of equations and then taking a limit.

The equations governing the behaviour of an incompressible micropolar fluid, in the absence of body force and body couple are

$$\varrho\left(\frac{\partial v}{\partial t} + (v \cdot \nabla)\, v\right) = (\mu + \varkappa)\, \nabla^2 v + \varkappa \nabla \times \nu - \nabla p\,, \tag{1}$$

$$\varrho j\left(\frac{\partial \nu}{\partial t} + (v \cdot \nabla)\, \nu\right) = (\alpha + \beta)\, \nabla\nabla \cdot \nu + \gamma \nabla^2 \nu + \varkappa \nabla \times v - 2\varkappa \nu\,, \tag{2}$$

$$\nabla \cdot v = 0\,, \tag{3}$$

where v is the velocity, ν the microrotation, p the thermodynamic pressure, ϱ the density and j the micro-inertia; $\mu, \varkappa, \alpha, \beta$ and γ are the material constants.

Here v, ν are vector-functions of x in Ω, a domain in R^3, and t. At this time, it is convenient to treat v and ν as functions of t alone, with values in one of the function spaces defined below. Accordingly, we write $v: t \to v(t)$, and $\nu: t \to \nu(t)$ for the velocity and microrotation of a flow, and we write the equations (1)—(3) in the form

$$\dot v(t) + (v(t) \cdot \nabla)\, v(t) = (\mu + \varkappa)\, \nabla^2 v(t) + \varkappa \nabla \times \nu(t) - \frac{1}{\varrho} \nabla p(t)\,, \tag{4}$$

$$j\dot\nu(t) + j(v(t) \cdot \nabla)\, \nu(t) = (\alpha + \beta)\, \nabla\nabla \cdot \nu(t) + \gamma \nabla^2 \nu(t) + \varkappa \nabla \times v(t) - 2\varkappa \nu(t)\,, \tag{5}$$

$$\nabla \cdot v(t) = 0\,. \tag{6}$$

Equations (4)—(6) are valid for $t > 0$ (and for x in the domain Ω of the fluid). In addition to (4)—(6) we want v and ν to satisfy

$$v(t) = 0 \quad \text{for} \quad x \in \partial\Omega\,, \qquad \nu(t) = 0 \quad \text{for} \quad x \in \partial\Omega\,, \tag{7}$$

and

$$v(0) = v^0\,, \qquad \nu(0) = \nu^0 \tag{8}$$

where $\partial\Omega$ denotes the boundary of Ω.

Definition. *A pair of functions $\{v, \nu\}: t \to \{v(t), \nu(t)\}$ is called a weak solution of the initial-boundary value problem* (4)—(8) *if it belongs to* $L^2((0, \infty); \mathcal{H}_0^{1,2} \times H_0^{1,2})$ and satisfies

$$\begin{gathered}\int_0^\infty [(v(t), \dot u(t)) + (v(t) \cdot \nabla u(t), v(t)) + (\mu + \varkappa)\, (v(t), \nabla^2 u(t)) + \varkappa(\nu(t), \nabla \times u(t))]\, dt = -(v^\circ, u(0))\,, \\ \int_0^\infty [j(\nu(t), \dot\psi(t)) + j(v(t) \cdot \nabla\psi(t), \nu(t)) + (\alpha + \beta)\, (\nu(t), \nabla\nabla \cdot \psi(t)) + \\ + \gamma(\nu(t), \nabla^2\psi(t)) + \varkappa(v(t), \nabla \times \psi(t)) - 2\varkappa(\nu(t), \psi(t))]\, dt = -j(\nu^0, \psi(0))\end{gathered} \tag{9}$$

for all $u \in C_0^1([0, \infty); \mathcal{C}_0^\infty)$ *and* $\psi \in C_0^1([0, \infty); C_0^\infty)$.

Here C_0^∞ is the set of vector functions defined in Ω, having continuous derivatives of all orders and being zero in a neighbourhood of $\partial\Omega$, $\mathscr{C}_0^\infty$ is the set of vector-functions in C_0^∞ that are divergence free; $H_0^{1,2}$ and $\mathscr{H}_0^{1,2}$ are the completions of C_0^∞ and $\mathscr{C}_0^\infty$, respectively, in the norm

$$||u||_H = ||\nabla u||_2 \equiv \Big(\int_\Omega |\nabla u|^2 \, dx\Big)^{1/2} .$$

The space $C_0^1([0, \infty); \mathscr{C}_0^\infty)$ is the set of all functions $\varphi: t \to \varphi(t)$ defined and with continuous derivative for all t from $[0, \infty)$, with φ taking values in $\mathscr{C}_0^\infty$.

In the rest of paper we shall intensively use the results of chapter 10 of [6].

2. The quantized equations

If, in (4), (5), we replace terms $\dot v(t)$ and $\dot\nu(t)$ by approximations, of the form $(1/h)\,[v(t) - v(t-h)]$ and $(1/h)\,[\nu(t) - \nu(t-h)]$, respectively, we may resaonably expect any solutions of the new equations to be close to a solution of (4)—(6) when h is small. The same thing should be true if we replace the terms $(v(t)\cdot\nabla)\,v(t)$ and $(v(t)\cdot\nabla)\,\nu(t)$ by $(v(t-h), \nabla)\,v(t)$ and $(v(t-h)\cdot\nabla)\,\nu(t)$, respectively.

The equations obtained from (4)—(6) after these replacements are

$$\begin{aligned} &v(t) - v(t-h) + h(v(t-h)\cdot\nabla)\,v(t) - h(\mu+\varkappa)\,\nabla^2 v(t) - \varkappa h\nabla\times\nu(t) = -\frac{h}{\varrho}\nabla p(t)\,,\\ &j\nu(t) - j\nu(t-h) + jh(v(t-h)\cdot\nabla)\,\nu(t) - h(\alpha+\beta)\,\nabla\nabla\cdot\nu(t) - \\ &\quad - \gamma h\nabla^2\nu(t) - \varkappa h\nabla\times v(t) + 2\varkappa h\nu(t) = 0\,,\\ &\nabla\cdot v(t) = 0\,. \end{aligned} \tag{10}$$

$v(0)$ and $\nu(0)$ are given. Setting $t = h$ in (10) we obtain equations for $\{v(h), \nu(h)\}$. If they can be solved, $\{v(h), \nu(h)\}$ can then be found. Then, (10) with $t = 2h$ become equations for $\{v(2h), \nu(2h)\}$, since $\{v(h), \nu(h)\}$ are known. Again these equations can, hopefully, be solved for $\{v(2h), \nu(2h)\}$. The process can clearly be continued and, in this way, we can solve successively for $\{v(h), \nu(h)\}$, $\{v(2h), \nu(2h)\}$, ..., $\{v(kh), \nu(kh)\}$, ..., given that $\{v(0), \nu(0)\}$ is known.

Write

$$\begin{aligned} v^k &= v(kh), \qquad k = 0, 1, \dots,\\ \nu^k &= \nu(kh), \qquad k = 0, 1, \dots, \end{aligned} \tag{11}$$

and

$$p^k = p(kh), \qquad k = 0, 1, \dots .$$

Then, setting $t = kh$ in (4)—(6), we obtain the following equations for $\{v^k, \nu^k\}$:

$$\begin{aligned} &v^k - v^{k-1} + hv^{k-1}\cdot\nabla v^k - h(\mu+\varkappa)\,\nabla^2 v^k - kh\nabla\times\nu^k = -\frac{h}{\varrho}\nabla p^k\,,\\ &j\nu^k - j\nu^{k-1} + jhv^{k-1}\cdot\nabla\nu^k - h(\alpha+\beta)\,\nabla\nabla\cdot\nu^k - \gamma h\nabla^2\nu^k - \varkappa h\nabla\times v^k + 2\varkappa h\nu^k = 0\,, \end{aligned} \tag{12}$$

$$\nabla\cdot v^k = 0\,, \tag{13}$$

$k = 1, 2, \dots$. We call (12) and (13) the "quantized equations".

Equations (12) and (13) have the form

$$\begin{aligned} &v + h\omega\cdot\nabla v - h(\mu+\varkappa)\,\nabla^2 v - \varkappa h\,\nabla\times\nu = \omega - \frac{h}{\varrho}\nabla p\,,\\ &j\nu + jh\omega\cdot\nabla\nu - h(\alpha+\beta)\,\nabla\nabla\cdot\nu - \gamma h\nabla^2\nu - \varkappa h\nabla\times v + 2\varkappa h\nu = j\chi\,, \end{aligned} \tag{14}$$

$$\nabla\cdot v = 0 \tag{15}$$

where we have written ω for v^{k-1}, χ for ν^{k-1}, v for v^k and ν for ν^k. When ω and χ are given, we call $\{v, \nu\}$ a weak solution of (14) and (15) if

$$\{v, \nu\} \in \mathscr{H}_0^{1,2} \times H_0^{1,2}\,, \tag{16}$$

and

$$\begin{aligned} &(v, u) + h(\omega\cdot\nabla v, u) + h(\mu+\varkappa)\,(\nabla v, \nabla u) - \varkappa h(\nu, \nabla\times u) = (\omega, u)\,,\\ &j(\nu, \psi) + jh(\omega\cdot\nabla\nu, \psi) + h(\alpha+\beta)\,(\nabla\cdot\nu, \nabla\cdot\psi) + \gamma h(\nabla\nu, \nabla\psi) - \\ &\quad - \varkappa h(\nabla\times v, \psi) + 2\varkappa h(\nu, \psi) = j(\chi, \psi)\,, \end{aligned} \tag{17}$$

for all $u \in \mathscr{C}_0^\infty$ and $\psi \in \mathscr{C}_0^\infty$ ($\omega \in \mathscr{L}^2$, $\chi \in L^2$).

Lemma 1. *Let Ω be bounded, and let $h, \mu, \varkappa, \gamma, \alpha + \beta > 0$. Then,* (14) *and* (15) *have a weak solution for all $\omega \in \mathscr{L}^2$ and $\chi \in L^2$.*

Proof. We first prove the lemma assuming that $\omega \in \mathscr{E}_0^\infty$. Let $\{v, \nu\}$ be any element of $\mathscr{H}_0^{1,2} \times H_0^{1,2}$, and define two linear functionals $E_{\{v,\nu\}}$ and $F_{\{v,\nu\}}$ by the formulas

$$\begin{aligned} E_{\{v,\nu\}}(u) &\equiv (v, u) + h(\omega \cdot \nabla v, u) + h(\mu + \varkappa)(\nabla v, \nabla u) - \varkappa h(\nu, \nabla \times u)\,, \\ F_{\{v,\nu\}}(\psi) &\equiv j(\nu, \psi) + jh(\omega \cdot \nabla \nu, \psi) + h(\alpha + \beta)(\nabla \cdot \nu, \nabla \cdot \psi) + \gamma h(\nabla \nu, \nabla \psi) - \\ &- \varkappa h(\nabla \times v, \psi) + 2\varkappa h(\nu, \psi)\,. \end{aligned} \tag{18}$$

If $\omega \in \mathscr{E}_0^\infty$, $E_{\{v,\nu\}}$ and $F_{\{v,\nu\}}$ are bounded linear functionals on $\mathscr{H}_0^{1,2}$ and $H_0^{1,2}$, respectively.

$$|E_{\{v,\nu\}}(u)| \leqq c||\nabla u||_2\,, \qquad |F_{\{v,\nu\}}(\psi)| \leqq c\,||\nabla \psi||_2\,,$$

for some positive constants c. Therefore, we conclude from the RIESZ representation theorem that there are the elements $v_1 \in \mathscr{H}_0^{1,2}$ and $\nu_1 \in H_0^{1,2}$ such that

$$E_{\{v,\nu\}}(u) = (v_1, u)_H, \qquad F_{\{v,\nu\}}(\psi) = (\nu_1, \psi)_H\,. \tag{19}$$

We have proved that for every pair $\{v, \nu\} \in \mathscr{H}_0^{1,2} \times H_0^{1,2}$, there are elements $v_1 \in \mathscr{H}_0^{1,2}$ and $\nu_1 \in H_0^{1,2}$ such that the linear functionals $E_{\{v,\nu\}}$ and $F_{\{v,\nu\}}$ defined by (18) have the forms (19). Let us denote the transformations from $\{v, \nu\}$ to v_1 and ν_1, respectively by A and B; we write

$$v_1 = A(v, \nu), \qquad \nu_1 = B(v, \nu)\,.$$

It is easy to see that these transformations are linear operators on $\mathscr{H}_0^{1,2} \times H_0^{1,2}$ and

$$E_{\{v,\nu\}}(u) = (A(v, \nu), u)_H\,, \qquad F_{\{v,\nu\}}(\psi) = (B(v, \nu), \psi)_H\,. \tag{20}$$

Setting $u = A(v, \nu)$ and $\psi = B(v, \nu)$, we obtain

$$||A(v, \nu)||_H \leqq c||\{v, \nu\}||_H\,, \qquad ||B(v, \nu)||_H \leqq c\,||\{v, \nu\}||_H$$

where

$$||\{v, \nu\}||_H^2 = ||v||_H^2 + ||\nu||_H^2\,.$$

Therefore, A and B are bounded.

Let us consider the linear operator $U\colon \mathscr{H}_0^{1,2} \times H_0^{1,2} \to \mathscr{H}_0^{1,2} \times H_0^{1,2}$ defined by

$$U(v, \nu) \equiv \{A(v, \nu),\ B(v, \nu)\}\,. \tag{21}$$

It is clear that U is bounded and

$$\begin{aligned} (U(v, \nu), \{v, \nu\}) &= ||v||_2^2 + h\mu\,||\nabla v||_2^2 + j||\nu||_2^2 + h(\alpha + \beta)\,||\nabla \cdot \nu||_2^2 + \\ &+ \gamma h\,||\nabla \nu||_2^2 + \varkappa h||b||_2^2 \geqq ch\,||\{v, \nu\}||_H^2 \end{aligned}$$

where b is the matrix of components $b_{kl} = (\partial/\partial x_l)\, v_k - \varepsilon_{ilk}\nu_i$ and v_k, ν_k are the components of the vectors v and ν, respectively; ε_{ilk} is the alternant tensor.

Since the left side of this inequality is the scalar product in $\mathscr{H}_0^{1,2} \times H_0^{1,2}$ of $U(v, \nu)$ and $\{v, \nu\}$, the LAX-MILGRAM lemma shows that the range of U is all of $\mathscr{H}_0^{1,2} \times H_0^{1,2}$.

The functionals

$$\mathscr{T}_1(u) = (\omega, u)\,, \qquad \mathscr{T}_2(\psi) = j(\chi, \psi)$$

are bounded on $\mathscr{H}_0^{1,2}$ and $H_0^{1,2}$, respectively. Therefore, again by the RIESZ representation theorem there are two elements $F \in \mathscr{H}_0^{1,2}$ and $L \in H_0^{1,2}$ such that

$$\mathscr{T}_1(u) = (F, u)_H\,, \qquad \mathscr{T}_2(\psi) = (L, \psi)_H\,,$$

for all $u \in \mathscr{H}_0^{1,2}$ and $\psi \in H_0^{1,2}$. The range of U is all $\mathscr{H}_0^{1,2} \times H_0^{1,2}$ and $\{F, L\} \in \mathscr{H}_0^{1,2} \times H_0^{1,2}$. Consequently, there is an $\{v, \nu\} \in \mathscr{H}_0^{1,2} \times H_0^{1,2}$ such that

$$U(v, \nu) = \{F, L\}\,, \tag{22}$$

or with (21) one yields

$$A(v, \nu) = F, \qquad B(v, \nu) = L\,. \tag{23}$$

We show that this pair of functions $\{v, \nu\}$ is a weak solution of (14) and (15). We already know that $\{v, \nu\}$ satisfies (16). As for (17), let $u \in \mathscr{E}_0^\infty$ and $\psi \in C_0^\infty$. Then,

$$\begin{aligned} (\omega, u) &= \mathscr{T}_1(u) = (F, u)_H = \big(A(v, \nu), u\big)_H = E_{\{v,\nu\}}(u)\,, \\ j(\chi, \psi) &= \mathscr{T}_2(u) = (L, \psi)_H = \big(B(v, \nu), \psi\big)_H = F_{\{v,\nu\}}(\psi)\,. \end{aligned}$$

From the definition of $E_{\{v,\nu\}}$ and $F_{\{v,\nu\}}$ these relations are just (17). The proof that $\{v, \nu\}$ is a weak solution of (14) and (15) is complete.

Now let $\omega \in \mathscr{L}^2$. Since $\mathscr{E}_2^\infty$ is dense in $\mathscr{L}^2$, there is a sequence $\{\omega_i\}_i$ of elements of $\mathscr{E}_0^\infty$ converging to ω in $\mathscr{L}^2$. Let $(\{v_i, \nu_i\})$ be the corresponding sequence of weak solutions of (14) and (15) whose existence has already been proved. Thus, $(\{v_i, \nu_i\}) \in \mathscr{H}_0^{1,2} \times H_0^{1,2}$ and each $\{v_i, \nu_i\}$ satisfies

$$\begin{aligned}&(v_i, u) + h(\omega_i \cdot \nabla v_i, u) + h(\mu + \varkappa)(\nabla v_i, \nabla u) - \varkappa h(\nu_i, \nabla \times u) = (\omega_i, u)\,,\\ &j(\nu_i, \psi) + jh(\omega_i \cdot \nabla \nu_i, \psi) + h(\alpha + \beta)(\nabla \cdot \nu_i, \nabla \cdot \psi) + \gamma h(\nabla \nu_i, \nabla \psi) -\\ &- \varkappa h(\nabla \times v_i, \psi) + 2\varkappa h(\nu_i, \psi) = j(\chi, \psi)\,,\end{aligned} \tag{24}$$

for all $u \in \mathscr{H}_0^{1,2}$ and $\psi \in H_0^{1,2}$. In these formulas, set $u = v_i$ and $\psi = \nu_i$. After obvious calculations we get

$$||v_i||_2^2 + j||\nu_i||_2^2 + 2h\mu\,||\nabla v_i||_2^2 + 2h(\alpha + \beta)||\nabla \cdot \nu_i||_2^2 + \gamma h\,||\nabla \nu_i||_2^2 + 2\varkappa h\,||b_i||_2^2 \leqq ||\omega_i||_2^2 + j||\chi||_2^2\,. \tag{25}$$

This shows that sequences $\{v_i\}$ and $\{\nu_i\}$ are bounded in $\mathscr{H}_0^{1,2}$ and $H_0^{1,2}$ respectively. Therefore, there are two subsequences converging weakly in $\mathscr{H}_0^{1,2}$ and $H_0^{1,2}$ respectively. In order not to proliferate subscripts, we call these subsequences $\{v_i\}$, $\{\nu_i\}$ again, and we call their weak limits v and ν respectively. From the fact that $\mathscr{H}_0^{1,2} \subset \mathscr{L}^2$ and $H_0^{1,2} \subset L^2$, with continuous injections, it follows the weak convergence of v_i in $\mathscr{L}^2$ and ν_i in L^2, and to the same limits. From the definition of scalar product in $\mathscr{L}^2 \times L^2$ we obtain that $\{v_i, \nu_i\}$ is weak convergent to $\{v, \nu\}$ in $\mathscr{L}^2 \times L^2$.

We also have weak convergence of $\{v_i, \nu_i\}$ in $\mathscr{H}_0^{1,2} \times H_0^{1,2}$. Consider next

$$\begin{aligned}&|(\omega_i \cdot \nabla v_i, u) - (\omega \cdot \nabla v, u)| \leqq ||\omega_i - \omega||_2\,||\nabla v_i||_2\,||u||_\infty + |(\omega \cdot \nabla(v_i - v), u)|\,,\\ &|(\omega_i \cdot \nabla \nu_i, \psi) - (\omega \cdot \nabla \nu, \psi)| \leqq ||\omega_i - \omega||_2\,||\nabla \nu_i||_2\,||\psi||_\infty + |(\omega \cdot \nabla(\nu_i - \nu), \psi)|\,.\end{aligned} \tag{26}$$

$||\nabla v_i||_2$ and $||\nabla \nu_i||_2$ are bonnded. Also, $\{\omega_i\}$ converges (strongly) to ω in $\mathscr{L}^2$. Therefore the first term of the right of $(26)_1$ and $(26)_2$ goes to zero as i goes to infinity. Any product of components of ω and u, is square integrable, since $\omega \in \mathscr{L}^2$ and $u \in \mathscr{C}_0^\infty \subset L^\infty$. Since $\{v_i\}$ converges weakly to v in $\mathscr{H}_0^{1,2}$ this implies that the second term on the right of $(26)_1$ goes to zero. In the same way we deduce that the second term on the right of $(26)_2$ goes to zero also. Therefore

$$\begin{aligned}&(\omega_i \cdot \nabla v_i, u) \to (\omega \cdot \nabla v, u)\,,\\ &(\omega_i \cdot \nabla \nu_i, \psi) \to (\omega \cdot \nabla \nu, \psi), \quad \text{as} \quad i \to \infty\,.\end{aligned}$$

Putting the results together and letting i go to infinity in (24), we obtain (17). This shows that $\{v, \nu\}$ is a weak solution of (14) and (15) and complete the proof of lemma 1.

Lemma 2. *Let Ω be bounded and let $h, \mu, \varkappa, \gamma, \alpha + \beta > 0$. Let $\omega \in \mathscr{L}^2$, $\chi \in L^2$, and let $\{v, \nu\}$ be the corresponding weak solution of* (14) *and* (15). *Then*

$$\begin{aligned}&||v||_2^2 + j||\nu||_2^2 + 2h\mu\,||\nabla v||_2^2 + 2h(\alpha + \beta)\,||\nabla \cdot \nu||_2^2 +\\ &+ 2\varkappa h\,||b||_2^2 + 2\gamma h\,||\nabla \nu||_2^2 \leqq ||\omega||_2^2 + j||\chi||_2^2\,,\end{aligned} \tag{27}$$

$$\begin{aligned}&||v||_2^2 + j||\nu||_2^2 + ||v - \omega||_2^2 + j||\nu - \chi||_2^2 + 2h\mu\,||\nabla v||_2^2 + 2\gamma h\,||\nabla \nu||_2^2 +\\ &+ 2h(\alpha + \beta)\,||\nabla \cdot \nu||_2^2 + 2\varkappa h||b||_2^2 \leqq ||\omega||_2^2 + j||\chi||_2^2\,.\end{aligned} \tag{28}$$

Proof. As we saw in the proof of lemma 1, we may assume there is a sequence $\{\omega_i\}$ of elements of $\mathscr{C}_0^\infty$ converging strongly to ω in $\mathscr{L}^2$, and a corresponding sequence $(\{v_i, \nu_i\})_i$ of weak solution of (14) and (15) converging weakly to $\{v, \nu\}$ in both $\mathscr{H}_0^{1,2} \times H_0^{1,2}$ and $\mathscr{L}^2 \times L^2$. Then (25) is satisfied for all i. If, in (25), we allow i to go to infinity, the right side goes to $||\omega||_2^2 + j||\chi||_2^2$, while the left side is never less than

$$||v||_2^2 + j||\nu||_2^2 + 2h\mu\,||\nabla v||_2^2 + 2h(\alpha + \beta)\,||\nabla \cdot \nu||_2^2 + 2\gamma h\,||\nabla \nu||_2^2 + 2\varkappa h\,||b||_2^2\,.$$

Thus, we obtain (27). If in (24) set $u = v_i$, and $\psi = \nu_i$ we get, after obvious calculus

$$\begin{aligned}&||v_i||_2^2 + j||\nu_i||_2^2 + ||v_i - \omega_i||_2^2 + j||\nu_i - \chi||_2^2 + 2h\mu\,||\nabla v_i||_2^2 + 2h(\alpha + \beta)\,||\nabla \cdot \nu_i||_2^2 +\\ &+ 2\gamma h|\,|\nabla \nu_i||_2^2 + 2\varkappa h||b_i||_2^2 \leqq ||\omega_i||_2^2 + j||\chi||_2^2\,.\end{aligned}$$

For $i \to \infty$ this formula gives (28).

Now, we can prove the main results of this section.

Theorem 1. *Let Ω be bounded, and let $h, \mu, \alpha + \beta, \varkappa, \gamma > 0$. If $\{v^0, \nu^0\} \in \mathscr{L}^2 \times L^2$, the quantized equations* (12) *and* (13) *have a solution $\{v^k, \nu^k\}$ for all $k = 1, 2, \ldots$ This solution satisfies*

$$\begin{aligned}&||v^k||_2^2 + j||\nu^k||_2^2 + 2h\mu\,||\nabla v^k||_2^2 + 2h(\alpha + \beta)\,||\nabla \cdot \nu^k||_2^2 +\\ &+ 2\gamma h\,||\nabla \nu^k||_2^2 + 2\varkappa h\,||b^k||_2^2 \leqq ||v^0||_2^2 + j||\nu^0||_2^2\,,\end{aligned} \tag{29}$$

$$\begin{aligned}&\sum_{k=1}^{\infty} [||v^k - v^{k-1}||_2^2 + j||\nu^k - \nu^{k-1}||_2^2] +\\ &+ 2h \sum_{L=1}^{\infty} [\mu\,||\nabla v^k||_2^2 + (\alpha + \beta)\,||\nabla \cdot \nu^k||_2^2 + \gamma\,||\nabla \nu^k||_2^2 + \varkappa\,||b^k||_2^2] \leqq ||v^0||_2^2 + j||\nu^0||_2^2\,,\end{aligned} \tag{30}$$

where b^k is the matrix of components $b_{ij}^k = (\partial/\partial x_j)\, v_i^k - \varepsilon_{mji} \nu_m^k$.

Proof. That the sequence $(\{v^k, \nu^k\})$ of solutions exists is just lemma 1, for setting $\omega = v^{k-1}$, $\chi = \nu^{k-1}$, $v = v^k$ and $\nu = \nu^k$ in (14) and (15) reduces then to (12) and (13). Thus it remains to prove (29) and (30). Setting $\omega = v^{k-1}$, $\chi = \nu^{k-1}$, $v = v^k$ and $\nu = \nu^k$ in (27), and using an inductive argument we get (29). In the same way we get from (28) the relation (30).

3. Existence of solution of problem (4)—(8)

Let h be positive, and $\{v^0, \nu^0\}$ be any element of $\mathcal{L}^2 \times L^2$. Let $(\{v^k, \nu^k\})$ be the sequence of solutions of the quantized equations whose existence was just proved. Define the functions $v(t; h)$ and $\nu(t; h)$ as follows:

$$\left.\begin{aligned} v(t; h) &= v^k \\ \nu(t; h) &= \nu^k \end{aligned}\right\} \quad \text{for } \varkappa h \leqq t < (\varkappa + 1)\, h, \qquad (\varkappa = 0, 1, \ldots)\,. \tag{31}$$

In the rest of this note we assume that

$$\mu > 0\,, \quad \varkappa > 0\,, \quad \alpha + \beta > 0\,, \quad \gamma > 0$$

hold.

Lemma 3. *Let Ω be bounded and let $\{v^0, \nu^0\} \in \mathcal{L}^2 \times L^2$. Then, the functions $v(t, h)$ and $\nu(t; h)$ defined by* (31) *satisfy*

$$\begin{aligned} &\|v(t; h)\|_2^2 + j\|\nu(t; h)\|_2^2 + 2h\mu\|\nabla v(t; h)\|_2^2 + 2h(\alpha + \beta)\,\|\nabla \cdot \nu(t; h)\|_2^2 + \\ &+ 2\gamma h\,\|\nabla \nu(t; h)\|_2^2 + 2\varkappa h\|b(t; h)\|_2^2 \leqq \|v^0\|_2^2 + j\|\nu^0\|_2^2\,, \end{aligned} \tag{32}$$

for $t \geqq h$, and

$$\begin{aligned} &\frac{1}{h}\int_h^\infty [\|v(t; h) - v(t - h; h)\|_2^2 + j\,\|\nu(t; h) - \nu(t - h; h)\|_2^2]\, dt + \\ &+ 2\int_h^\infty [\mu\,\|\nabla v(t; h)\|_2^2 + (\alpha + \beta)\|\,\nabla \cdot \nu(t; h)\|_2^2 + \gamma\,\|\nabla \nu(t; h)\|_2^2 + \\ &+ \varkappa\|b(t; h)\|_2^2]\, dt \leqq \|v^0\|_2^2 + j\|\nu^0\|_2^2\,. \end{aligned} \tag{33}$$

If $u \in C_0^1([0, \infty); \mathscr{C}_0^\infty)$ and $\psi \in C_0^1([0, \infty); C_0^\infty)$ then

$$\begin{aligned} &\int_0^\infty [(v(t; h), \dot u(t)) + (v(t; h) \cdot \nabla u(t), v(t; h)) + (\mu + \varkappa)\,(v(t; h), \nabla^2 u(t)) + \\ &+ \varkappa(\nu(t; h), \nabla \times u(t))]\, dt = -\,(v^0, u(0)) + o(h^{1/2})\,, \\ &\int_0^\infty [j(\nu(t; h), \dot\psi(t)) + j(v(t; h) \cdot \nabla \psi(t), \nu(t; h)) + (\alpha + \beta)\,(\nu(t; h), \nabla\nabla \cdot \psi(t)) + \\ &+ \gamma(\nu(t; h), \nabla^2 \psi(t)) + \varkappa(\nabla \times v(t; h), \psi(t)) - 2\varkappa(\nu(t; h), \psi(t))]\, dt = -\,j(\nu^0, \psi(0)) + o(h^{1/2}) \end{aligned} \tag{34}$$

as $h \to 0$.

Proof. (32) is just (29), of course, rewritten in terms of $v(t; h)$ and $\nu(t; h)$. (33) is obtained from (30).

It remains to prove (34). Using the definition (31) of $v(t, h)$ and $\nu(t, h)$, and the fact that $\{v^k, \nu^k\}$ satisfies (17) we get

$$\begin{aligned} &(v(t; h) - v(t - h; h), u) - h(v(t - h; h) \cdot \nabla u, v(t; h)) - \\ &- h(\mu + \varkappa)\,(v(t; h), \nabla^2 u) - \varkappa h(\nu(t; h), \nabla \times u) = 0\,, \\ &j(\nu(t; h) - \nu(t - h; h), \psi) - hj(v(t - h, h) \cdot \nabla \psi, \nu(t; h)) - h(\alpha + \beta)\,(\nu(t; h), \nabla\nabla \cdot \psi) - \\ &- \gamma h(\nu(t; h), \nabla^2 \psi) + 2\varkappa h(\nu(t; h), \psi) - \varkappa h(\nabla \times v(t; h), \psi) = 0\,, \end{aligned} \tag{35}$$

for $t \geqq h$. If $u \in C_0^1([0, \infty); \mathscr{C}_0^\infty)$ and $\psi \in C_0^1([0, \infty); C_0^\infty)$ then we may replace u by $u(t)$, and ψ by $\psi(t)$ in (35). Integration with respect to t, followed by a little manipulation, there leads to

$$\begin{aligned} &\int_0^\infty \alpha(h; t)\, dt + (v^0, u(0)) = (v^0, u(0)) - \frac{1}{h}\int_0^h (v(t; h), u(t + h))\, dt + \int_0^h \alpha(h; t)\, dt + \\ &+ \int_h^\infty \left(v(t; h), \dot u(t) - \frac{u(t) - u(t + h)}{h}\right) dt + \int_h^\infty \big([v(t; h) - v(t - h; h)] \cdot \nabla u(t), v(t; h)\big)\, dt\,, \\ &\int_0^\infty \beta(h; t)\, dt + (j(\nu^0, \psi(0)) = j(\nu^0, \psi(0)) - \frac{1}{h}\int_0^h (\nu(t; h), \psi(t + h))\, dt + \int_0^h \beta(h; t)\, dt + \\ &+ j\int_0^\infty \left(\nu(t; h), \dot\psi(t) + \frac{\psi(t) - \psi(t + h)}{h}\right) dt + \int_h^\infty \big([v(t; h) - v(t - h; h)] \cdot \nabla \psi(t), \nu(t; h)\big)\, dt\,, \end{aligned} \tag{36}$$

where $\alpha(h, t)$, $\beta(h, t)$ denote the expressions under the integrals of first members of (34).

We estimate the eight terms of the right sides of (36) separately as in [6, p. 166] and we shall obtain (34). This completes the proof of lemma 3.

Lemma 4. *Let Ω be bounded, and let $\{v^0, \nu^0\} \in \mathcal{L}^2 \times L^2$. The functions $v(t; h)$ and $\nu(t; h)$ defined in (31) are weak continuously and satisfy*

$$|(v(t_2; h) - v(t_1; h), u)| \leqq c(|t_2 - t_1| + h),$$
$$|(\nu(t_2; h) - \nu(t_1; h), \psi)| \leqq c(|t_2 - t_1| + h),$$

for all $t_1, t_2 > 0$ and all $\{u, \psi\} \in \mathcal{C}_0^\infty \times C_0^\infty$. Here c are constants depending only on v^0, ν^0 and u or ψ.

The proof of this lemma is like that of lemma 11.4.2 of [6].

We can now prove the existence of a weak solution of problem (4)—(8).

Theorem 2. *Let Ω be bounded, and $\{v^0, \nu^0\} \in \mathcal{L}^2 \times L^2$. Then the initial-boundary value problem (4)—(8) has a weak solution.*

Proof. Because of (31), from (29) we have

$$||v(t; h)||_2^2 + j||\nu(t; h)||_2^2 \leqq ||v_0||_2^2 + j||\nu_0||_2^2, \tag{37}$$

for all $t \geqq 0$. Therefore, the families of functions $\{v(t; h)\}$ and $\{\nu(t; h)\}$ are bounded in $\mathcal{L}^2$ and L^2, respectively, for all $t \geqq 0$. It follows that for any fixed $t \geqq 0$, it is possible to find a sequence $\{h_n\}$, decreasing to zero, such that $\{v(t; h_n)\}$ converges weakly in $\mathcal{L}^2$, and $\{\nu(t; h_n)\}$ converges weakly in L^2. Let $\{t_m\}$ be the sequence of non-negative rationals. For each m, then, there is a sequence $\{h_{mn}\}$ such that $\{h_{mn}\}$ is a subsequence of $\{h_{m-1,n}\}$ while $\{v(t; h_{mn})\}$ and $\{\nu(t; h_{mn})\}$ converge weakly in $\mathcal{L}^2$ and L^2, respectively. Then, the usual diagonal argument shows that $\{v(t; h_{nn})\}$ and $\{\nu(t; h_{nn})\}$ converge weakly whenever t is rational. To keep the notation simple, we write $h_n = h_{nn}$.

By using lemma 4 we can proof that $\{v(t; h_n)\}$ and $\{\nu(t; h_n)\}$ converge weakly in $\mathcal{L}^2$ and L^2, respectively, for all $t \geqq 0$.

Let $v(t)$ and $\nu(t)$ be the weak limits of $\{v(t; h_n)\}$ and $\{\nu(t; h_n)\}$ in $\mathcal{L}^2$ and L^2 respectively. We remark that

$$||v(t)||_2^2 \leqq ||v^0||_2^2 + j||\nu^0||_2^2, \qquad ||\nu(t)||_2^2 \leqq ||v^0||_2^2 + j||\nu^0||_2^2, \qquad t \geqq 0. \tag{38}$$

This follows from (37).

By (37) we have

$$\int_0^T ||v(t; h_n)||_2^2\, dt \leqq T(||v^0||_2^2 + j||\nu^0||_2^2), \qquad \int_0^T ||\nu(t; h_n)||_2^2\, dt \leqq T\,(||v^0||_2^2 + j||\nu^0||_2^2),$$

for any $T > 0$. This shows that the sequences $\{v(t; h_n)\}$ and $\{\nu(t; h_n)\}$ are bounded in $L^2((0, T); \mathcal{L}^2)$ and $L^2((0, T); L^2)$ respectively. Therefore, there are subsequences of $\{v(t; h_n)\}$ and $\{\nu(t; h_n)\}$ that converge weakly in $L^2((0, T); L^2)$. We call these subsequences $\{v(t; h_n)\}$ and $\{\nu(t; h_n)\}$ again.

Next (33) shows that

$$\int_{h_n}^\infty ||\nabla v(t; h_n)||_2^2\, dt \leqq \frac{1}{2\mu}(||v^0||_2^2 + j||\nu^0||_2^2), \qquad \int_{h_n}^\infty ||\nabla \nu(t; h_n)||_2^2\, dt \leqq \frac{1}{2\gamma}(||v^0||_2^2 + j||\nu^0||_2^2).$$

Take $\delta > 0$. Since $\{h_n\}$ goes to zero as n goes to infinity, $h_n \leqq \delta$ if n is large enough. Therefore for n large enough

$$\int_\delta^T ||\nabla v(t; h_n)||_2^2\, dt \leqq \frac{1}{2\mu}(||v^0||_2^2 + j||\nu^0||_2^2), \qquad \int_\delta^T ||\nabla \nu(t; h_n)||_2^2\, dt \leqq \frac{1}{2\gamma}(||v^0||_2^2 + j||\nu^0||_2^2). \tag{39}$$

This shows that $\{v(t; h_n)\}$ and $\{\nu(t; h_n)\}$ are bounded in $L^2((\delta, T); H_0^{1,2})$ for any δ and T with $0 < \delta < T < \infty$. Therefore, as before, $\{v(t; h_n)\}$ and $\{\nu(t; h_n)\}$ contain subsequences converging weakly in $L^2((\delta, T); H_0^{1,2})$. If, again, we call these new subsequences $\{v(t; h_n)\}$ and $\{\nu(t; h_n)\}$, respectively, we see that $\{v(t; h_n)\}$ and $\{\nu(t, h_n)\}$ converge weakly in $L^2((\delta, T); H_0^{1,2})$, in $L^2((0, T), L^2)$, and in L^2 for every fixed $t \geqq 0$. The weak limits of $\{v(t; h_n)\}$ and $\{\nu(t; h_n)\}$ in all these spaces are, of course, the same functions $v(t)$ and $\nu(t)$, respectively, defined before.

From the definition of product spaces we can conclude that the sequence $(\{v(t; h_n), \nu(t; h_n)\})$ is weakly convergent in $L^2((0, T); \mathcal{L}^2 \times L^2)$, in $L^2((\delta, T); \mathcal{H}_0^{1,2} \times H_0^{1,2})$ and in $\mathcal{L}^2 \times L^2$ for every fixed $t \geqq 0$.

Now, let us show that $\{v(t; h_n)\}$ and $\{\nu(t; h_n)\}$ converge strongly to $v(t)$ and $\nu(t)$, respectively in $L^2((0, T); L^2)$. Since $v(t; h_n) - v(t) \in \mathcal{H}_0^{1,2}$ and $\nu(t; h_n) - \nu(t) \in H_0^{1,2}$ is follows that by lemma 2.9 of chap. 10 of [6]),

$$||v(t; h_n) - v(t)||_2^2 \leqq \sum_{i=1}^{N_1} |(v(t; h_n) - v(t), \omega_i)|^2 + \varepsilon\, ||\nabla v(t; h_n) - \nabla v(t)||_2^2,$$

and

$$||\nu(t; h_n) - \nu(t)||_2^2 \leqq \sum_{i=1}^{N_2} |(\nu(t; h_n) - \nu(t), \eta_i)|^2 + \varepsilon\, ||\nabla \nu(t; h_n) - \nabla \nu(t)||_2^2,$$

for almost all $t \in (\delta, T)$, and for every $\varepsilon > 0$. Integration with respect to t, followed by a little manipulation, then leads to

$$\int_0^T \|v(t; h_n) - v(t)\|_2^2 \, dt \leqq \left(4\delta + \frac{2\varepsilon}{\mu}\right)(\|v^0\|_2^2 + j\|\nu^0\|_2^2) + \sum_{i=1}^{N_1} \int_\delta^T |(v(t; h_n) - v(t), \omega_i)|^2 \, dt ,$$

$$\int_0^T \|\nu(t; h_n) - \nu(t)\|_2^2 \, dt \leqq \left(4\delta + \frac{2\varepsilon}{\gamma}\right)(\|v^0\|_2^2 + j\|\nu^0\|_2^2) + \sum_{i=1}^{N_2} \int_\delta^T |(\nu(t; h_n) - \nu(t), \eta_i)|^2 \, dt .$$

The terms $\left(4\delta + \frac{2\varepsilon}{\mu}\right)(\|v^0\|_2^2 + j\,\|\nu^0\|^2)$ and $\left(4\delta + \frac{2\varepsilon}{\gamma}\right)(\|v^0\|_2^2 + j\,\|\nu^0\|_2^2)$ can be made as small as desired by choosing δ and ε small enough. Once δ and ε are fixed, N_1 and N_2 are also fixed, and the last terms on the right can be made as small as desired by taking n large enough. This shows that $\{v(t; h_n)\}$ and $\{\nu(t; h_n)\}$ converge strongly to $v(t)$ and $\nu(t)$ respectively, in $L^2((0, T); L^2)$. It follows that $(\{v(t; h_n), \nu(t; h_n)\})$ converge strongly to $\{v(t), \nu(t)\}$ in $L^2((0, T); \mathcal{L}^2 \times L^2)$ also.

Let $\{u, \psi\} \in C_0^1([0, \infty); \mathcal{C}_0^\infty \times C_0^\infty)$. Choose T so large that $\{u(t), \nu(t)\} = \{0, 0\}$ for $t > T$. Then the integrals in (34) are just integrals over the interval $0 \leqq t \leqq T$. In (34), we replace h by h_n and let n go to infinity. Then

$$\int_0^\infty [(v(t; h_n), \dot{u}(t)) + (\mu + \varkappa)(v(t; h_n), \nabla^2 u(t)) + \varkappa(\nu(t; h_n), \nabla \times u(t))] \, dt \to$$

$$\to \int_0^\infty [(v(t), \dot{u}(t)) + (\mu + \varkappa)(v(t), \nabla^2 u(t)) + \varkappa(\nu(t), \nabla \times u(t))] \, dt ,$$

$$\int_0^\infty [j(\nu(t; h_n), \dot{\psi}(t)) + (\alpha + \beta)(\nu(t; h_n), \nabla\nabla \cdot \psi(t)) + \gamma(\nu(t; h_n), \nabla^2\psi(t)) +$$
$$+ \varkappa(v(t; h_n), \nabla \times \psi(t)) - 2\varkappa(\nu(t; h_n), \psi(t))] \, dt \to$$
$$\to \int_0^\infty [\gamma(\nu(t), \dot{\psi}(t)) + (\alpha + \beta)(\nu(t), \nabla\nabla \cdot \psi(t)) + \gamma(\nu(t), \nabla^2\psi(t)) + \varkappa(v(t), \nabla \times \psi(t)) - 2\varkappa(\nu(t), \psi(t))] \, dt ,$$

since $\{v(t; h_n)\}$ and $\{\nu(t; h_n)\}$ converge weakly to $v(t)$ and $\nu(t)$, respectively, in $L^2((0, T); L^2)$.

Obvious calculations yield

$$\left| \int_0^\infty (v(t; h) \cdot \nabla u(t), v(t; h_n)) \, dt - \int_0^\infty (v(t) \cdot \nabla u(t), v(t)) \, dt \right| \leqq c T^{1/2} \left(\int_0^T \|v(t; h_n) - v(t)\|_2^2 \, dt \right)^{1/2} \to 0 ,$$

as $n \to \infty$, and

$$\left| \int_0^\infty (v(t; h_n) \cdot \nabla \psi(t), \nu(t; h_n)) \, dt - \int_0^\infty (v(t) \cdot \nabla \psi(t), \nu(t)) \, dt \right| \leqq$$
$$\leqq c T^{1/2} \left(\int_0^T \|v(t; h_n) - v(t)\|_2^2 \, dt - \int_0^T \|\nu(t; h_n) - \nu(t)\|_2^2 \, dt \right) \to 0 ,$$

as $n \to \infty$, since $\{v(t; h_n)\}$ and $\{\nu(t; h_n)\}$ converge strongly to $v(t)$ and $\nu(t)$, respectively, in $L^2((0, T); L^2)$. Thus, the left side of (34) converges to the left side of (9) as n goes to infinity. The right side of (34) clearly goes to $-(v^0, u(0))$ and $-j(\nu^0, \psi(0))$ respectively, and therefore $\{v(t), \nu(t)\}$ satisfies (9). Also, letting h_n tend to zero in (39) and then letting δ go to zero and T to infinity we find

$$\int_0^\infty (\|\nabla v(t)\|_2^2 + \|\nabla \nu(t)\|_2^2) \, dt \leqq c\,(\|v^0\|_2^2 + j\|\nu^0\|_2^2) .$$

Therefore $\{v, \nu\} \in L^2((0, \infty); \mathcal{H}_0^{1,2} \times H_0^{1,2})$. This, plus the fact that $\{v, \nu\}$ satisfies (9) shows that it is a weak solution and completes the proof of theorem 2.

The weak solution satisfies an energy inequality. We state the energy inequality as a theorem.

Theorem 3. *Let Ω bounded, and let $\{v^0, \nu^0\} \in \mathcal{L}^2 \times L^2$. Let $\mu, \varkappa, \alpha + \beta, \gamma > 0$. Then, the weak solution $\{v, \nu\}$ satisfies the energy inequality*

$$\|v(t)\|_2^2 + j\|\nu(t)\|_2^2 + 2\mu \int_0^t \|\nabla v(\tau)\|_2^2 \, d\tau + 2(\alpha + \beta) \int_0^t \|\nabla \cdot \nu(\tau)\|_2^2 \, d\tau +$$
$$+ 2\gamma \int_0^t \|\nabla \nu(\tau)\|_2^2 \, d\tau + 2\varkappa \int_0^t \|b(\tau)\|_2^2 \, d\tau \leqq \|v^0\|_2^2 + j\|\nu^0\|_2^2 . \tag{40}$$

Proof. Proving theorem 1 we showed that

$$\|v^k\|_2^2 + j\|\nu^k\|_2^2 + 2h\mu \|\nabla v^k\|_2^2 + 2h(\alpha + \beta) \|\nabla \cdot \nu^k\|_2^2 +$$
$$+ 2\gamma h \|\nabla \nu^k\|_2^2 + 2\varkappa h \|b^k\|_2^2 \leqq \|v^{k-1}\|_2^2 + j\|\nu^{k-1}\|_2^2 ,$$

where $(\{v^k, \nu^k\})$ is a sequence of weak solutions of the quantized equations. Summing on k, we find

$$||v^i||_2^2 + j||\nu^i||_2^2 + 2h\mu \sum_{k=1}^{i} ||\nabla v^k||_2^2 - 2h(\alpha + \beta) \sum_{k=1}^{i} ||\nabla \cdot \nu^k||_2^2 +$$
$$+ 2\gamma h \sum_{k=1}^{i} ||\nabla \nu^k||_2^2 + 2\varkappa h \sum_{k=1}^{i} ||b^k||_2^2 \leqq ||v^0||_2^2 + j||\nu^0||_2^2 .$$

Take $\delta > 0$. If $ih \leqq t < (i + 1) h$ and taking account of (31) the above relation yields

$$||v(t; h)||_2^2 + j||\nu(t; h)||_2^2 + 2\mu \int_\delta^t ||\nabla v(t; \tau)||_2^2 \, d\tau + 2(\alpha + \beta) \int_\delta^t ||\nabla \cdot \nu(t; \tau)||_2^2 \, d\tau +$$
$$+ 2\gamma \int_\delta^t ||\nabla \nu(t; \tau)||_2^2 \, d\tau + 2\varkappa \int_\delta^t ||b(t, \tau)||_2^2 \, d\tau \leqq ||v^0||_2^2 + j||\nu^0||_2^2 . \tag{41}$$

We saw in the proof of theorem 2 that there is a sequence $\{h_n\}$ such that sequences $\{v(t; h_n)\}$ and $\{\nu(t; h_n)\}$ convergence weakly to v and ν, respectively, the weak solution of the problem (4)—(8). Writing h_n for h in (41) letting n go to infinity, and then letting δ go to zero, we find (40). This completes the proof of theorem 3.

The weak solution of theorem 2 does not have much smoothness with respect to t, but it does have some. How much, is the content of our next result.

Theorem 4. *The weak solution $\{v, \nu\}$ of theorem 2 is continuous as a function of t in the weak topology of $\mathcal{L}^2 \times L^2$. Moreover it satisfies*

$$||v(t) - v^0||_2^2 + j||\nu(t) - \nu^0||_2^2 \to 0, \quad \text{as} \quad t \to 0 .$$

A proof of this theorem is given in [8].

The uniqueness of solution of the initial boundary-value problem for the equations which describe the flow of incompressible micropolar fluids is proved in [9].

References

1 ERINGEN, A. C., Theory of micropolar fluids, J. Math. Mech. **16** (1967), 1—18.
2 FÖRSTE, J., Zur Existenz stationärer Strömungen bei inkompressiblen mikropolaren Flüssigkeiten, Mber. Dt. Akad. Wiss. **13** (1971), 1—8.
3 FÖRSTE, J., Instationäre Strömung einer inkompressiblen mikropolaren Flüssigkeit, Mber. Dt. Akad. Wiss. **13** (1971), 741—748.
4 SAVA, V. A., On the existence and uniqueness of the solution of a boundary-value problem in the theory of incompressible micropolar fluids, An. Sti. Univ. „Al. I. Cuza" Iași, Sect. I Mat. **19** (1973), 449—460.
5 SAVA, V., Existence theorems in the theory of micropolar fluids (in press).
6 LANGE, H., Die Existenz von Lösungen der Gleichungen, welche die Strömung inkompressibler mikropolarer Flüssigkeiten beschreiben, ZAMM **56** (1976), 129—139.
7 SHINBROT, M., Lectures on Fluids Mechanics, Gordon and Breach, New York, 1973.
8 SAVA, V. A., Ph. D. Thesis, Univ. „Al. I. Cuza" Iași 1976.
9 SAVA, V. A., An uniqueness theorem for incompressible micropolar fluids, An. St. Univ. „Al. I. Cuza" Iași, sect. J. Mat. **24** (1978), 161—170.

Eingereicht am 17. 12. 1976

Adress: Dr. V. A. SAVA, Universitatea „Al. I. Cuza", Seminarul Matematic, R-6600 — Iași, Romania

KLEINE MITTEILUNGEN

ZAMM 58, 519 – 520 (1978)

J. Nieminen

Screens and Rounding Mappings

The purpose of this short paper is to extend Kulisch's results on screens for complete lattices and linear orders [5, 6] and to continue the extending process of Apostolatos [1]. At first we shall characterize upper and lower screens R, and consequently, screens as well, by means of rounding mappings induced by R. Thereafter we shall consider properties of screens being similar to screens of linear orders.

Let L be a lattice. $[a)$ and $(a]$ denote the *principal filter* and *ideal* of L, repectively, generated by an element $a \in L$. A subset $R \subset L$ is called a *lower screen* of L, if for any $a \in L$ there is an element $x \in (a] \cap R$ such that $x \geqq y$ for every $y \in (a] \cap R$, i.e. there is a greatest element in $(a] \cap R$ for any $a \in L$. R is called an *upper screen* of L, if $[a) \cap R$ contains a least element for any $a \in L$. R is a *screen* of L, if it is an upper and a lower screen of L.

Let R be an upper screen (or lower screen, or a screen) of L. A mapping $\beta\colon L \to R$ is called a *rounding* of L into R if $a \leqq b \Rightarrow \beta(a) \leqq \beta(b)$ for any two $a, b \in L$. A rounding β is called *optimal*, if $\beta(a) = a$ for any $a \in R$.

The least and greatest elements of $[a) \cap R$ and $(a] \cap R$, respectively, allows us to define two mappings on L characterizing upper and lower screens of L.

Theorem 1: *A non-empty subset R of a lattice L is an upper screen of L if and only if R is the set of all fixpoints of a single-valued mapping $\varphi\colon L \to R$ defined as $\varphi(a) = \inf\{[a) \cap R\}$ for every $a \in L$ and having the following properties:*

(i) $a \leqq \varphi(a)$ *for every* $a \in L$;

(ii) $a \leqq b \Rightarrow \varphi(a) \leqq \varphi(b)$ *for every two* $a, b \in L$;

(iii) $\varphi(\varphi(a)) = \varphi(a)$ *for every* $a \in L$;

(iv) $\varphi(a \wedge b) \leqq \varphi(a) \wedge \varphi(b)$ *for every two* $a, b \in L$.

Proof: Let R be an upper screen of L. According to the definition of an upper screen, the mapping $\varphi(a) = \inf\{[a) \cap R\}$ is single-valued. The definitions of φ and R imply that $a \leqq \varphi(a)$, and moreover, the elements of R are the only fixpoints of φ, whence $\varphi(\varphi(a)) = \varphi(a)$. As $a \leqq b$, $[a) \cap R \supset [b) \cap R$, whence $\inf\{[a) \cap R\} \leqq \inf\{[b) \cap R\}$ and $\varphi(a) \leqq \varphi(b)$. As $a \wedge b \leqq a, b$, (ii) implies that $\varphi(a \wedge b) \leqq \varphi(a), \varphi(b)$, and thus $\varphi(a \wedge b) \leqq \varphi(a) \wedge \varphi(b)$, too.

Conversely, let φ be a single-valued mapping on L having the properties (i)—(iv). According to (iii),

$$\Phi = \{x \mid x = \varphi(q) \text{ for some } q \in L\}$$

is the set of all fixpoints of φ. As $a \leqq \varphi(a)$, $[a) \cap \Phi \neq \emptyset$ for every $a \in L$. Moreover, let $\varphi(a), \varphi(b) \in [a) \cap \Phi$, whence $a \leqq \varphi(a), \varphi(b)$. Now

$$\varphi(a) = \varphi(a \wedge \varphi(b)) \leqq \varphi(a) \wedge \varphi(\varphi(b)) = \varphi(a) \wedge \varphi(b)\,,$$

and thus $\varphi(a) \leqq \varphi(b)$ for every $\varphi(b) \in [a) \cap \Phi$. Therefore Φ is an upper screen of L. This completes the proof.

The dual proof shows that a *lower screen R of L is characterized by a mapping $\lambda\colon L \to R$ having the properties*

(i′) $\lambda(a) \leqq a$ *for every* $a \in L$;

(ii′) $a \leqq b \Rightarrow \lambda(a) \leqq \lambda(b)$ *for every two* $a, b \in L$;

(iii′) $\lambda(\lambda(a)) = \lambda(a)$ *for every* $a \in L$;

(iv′) $\lambda(a \vee b) \geqq \lambda(a) \vee \lambda(b)$ *for every two* $a, b \in L$.

As a screen is a lower and an upper screen, *the mappings λ and φ characterize together screens of L.*

For screens one can prove

Theorem 2: *Let R be a screen of L. Then $\varphi(a \vee b) = \varphi(a) \vee \varphi(b)$ and $\lambda(a \wedge b) = \lambda(a) \wedge \lambda(b)$.*

Proof: We shall consider the mapping φ only; a dual proof shows the asserted property for λ.

$a \leqq \varphi(a)$ and $b \leqq \varphi(b)$, whence $a \vee b \leqq \varphi(a) \vee \varphi(b)$ and further, $\varphi(a \vee b) \leqq \varphi(\varphi(a) \vee \varphi(b))$. If $\varphi(a) \vee \varphi(b) \in R$, then $\varphi(\varphi(a) \vee \varphi(b)) = \varphi(a) \vee \varphi(b)$, and thus $\varphi(a \vee b) \leqq \varphi(a) \vee \varphi(b)$. On the other hand, $a, b \leqq a \vee b$, and so $\varphi(a) \vee \varphi(b) \leqq \varphi(a \vee b)$. By combining the above results,

$$\varphi(a \vee b) = \varphi(a) \vee \varphi(b) \quad \text{if} \quad \varphi(a) \vee \varphi(b) \in R\,.$$

Trivially, $\varphi(a), \varphi(b) \geqq \varphi(a) \wedge \varphi(b)$, and so

$$\varphi(a), \varphi(b) \in [\varphi(a) \wedge \varphi(b)) \cap R\,.$$

Thus $\varphi(\varphi(a) \wedge \varphi(b)) \leqq \varphi(a), \varphi(b)$, and consequently, $\varphi(\varphi(a) \wedge \varphi(b)) \leqq \varphi(a) \wedge \varphi(b)$. But as

$$\varphi(\varphi(a) \wedge \varphi(b)) \in [\varphi(a) \wedge \varphi(b)) \cap R\,,$$

$$\varphi(\varphi(a) \wedge \varphi(b)) = \varphi(a) \wedge \varphi(b) \in R\,.$$

The dual proof holds for λ, and as R is a screen, $\lambda(a) \vee \lambda(b) \in R$ for any two elements $\lambda(a), \lambda(b) \in R$. As $\varphi(a), \varphi(b) \in R$, also $\varphi(a) \vee \varphi(b) \in R$, and the theorem follows.

Theorem 2 shows that the rounding mapping φ on L is a join-homomorphism of L, and λ a meet-homomorphism, respectively.

In [5] Kulisch considered properties of screens when L was a linear order. In [5, Lemma 4] he proved that $\varphi(a) > q > \lambda(a)$ holds for no $q \in R$ and no $a \in L$ when L is linear, but this does not hold in general as one can easily see. In the following we shall look for properties of R which imply that $\varphi(a) > q > \lambda(a)$ holds for no $q \in R$ and no $a \in L$.

One can construct examples for showing that the property

$$\varphi(a \wedge b) = \varphi(a) \wedge \varphi(b)\,, \qquad \lambda(a \vee b) = \lambda(a) \vee \lambda(b)$$

is not equivalent with the desired property. A fruitful way is to construct a covering of L induced by R and consider the relation given by this covering.

Let R be a screen of a lattice L. We define a family $\mathcal{M}^*$ of sets generated by R in L. Let $q, p \in R$ and $q \leqq p$. If $q = p$, $M_{qp} = \{p\}$, and if $q < p$,

$$M_{qp} = \{t \mid t \in L, \varphi(t) = p, \lambda(t) = q, \text{ and } \varphi \text{ and } \lambda \text{ are determined by } R\} \cup \{q\} \cup \{p\}\,.$$

$\mathcal{M}$ is a subfamily of $\mathcal{M}^*$ such that $M_{qp} \in \mathcal{M}$ if and only if $M_{qp} \cap \{L \setminus R\} \neq \emptyset$ when $q < p$, and $\{p\} \in \mathcal{M}$ when $p = q$ and p does not belong to any other set of $\mathcal{M}$.

Lemma 1: *Let R be a screen of a lattice L and φ and λ the mappings characterizing R. There are no elements $b \in R$ such that $\varphi(a) > b > \lambda(a)$ for some $a \in L$ if and only if each subset $M_{qp} \in \mathcal{M}$ is a convex sublattice of L.*

Proof: Let each $M_{qp} \in \mathcal{M}$ be a convex sublattice of L, and let $a \in L$ be an element and $b \in R$ another such that $\varphi(a) > b > \lambda(a)$. As $\varphi(a) > \lambda(a)$, and as R is the set of fixpoints of φ and λ, $a \in L \setminus R$, whence $M_{\lambda(a)\varphi(a)} \in \mathcal{M}$. Further, $M_{\lambda(a)\varphi(a)}$ is convex, and thus $b \in M_{\lambda(a)\varphi(a)}$. As $b \in R$, $\varphi(b) = \lambda(b) = b \neq \varphi(a), \lambda(a)$, which shows that b does not satisfy the definition of $M_{\lambda(a)\varphi(a)}$. This is a contradiction, and therefore there is not an element $b \in R$ such that $\varphi(a) > b > \lambda(a)$ for some $a \in L$.

Conversely, we assume that there is no element $b \in R$ such that $\varphi(a) > b > \lambda(a)$ for some $a \in L$. We shall show that then the sets M_{qp} of $\mathcal{M}$ are convex sublattices of L. Let $d, c \in M_{qp}$, whence

1°: $\varphi(d) = \varphi(c) = p$ and $\lambda(d) = \lambda(c) = q$,

or

2°: $d, c \in \{p, q\}$.

1° According to Theorem 2, $\varphi(c \vee d) = p$ and $\lambda(c \wedge d) = q$. Let us consider $\varphi(c \wedge d)$. As $c, d \geqq q = \lambda(c) = \lambda(d)$, $c \wedge d \geqq q$, too. According to (ii), $\varphi(c \wedge d) \geqq q$, and according to (iv),

$$p = \varphi(c) \wedge \varphi(d) \geqq \varphi(c \wedge d)\,.$$

The following two possibilities exist: $\varphi(c \wedge d) = q$ or $\varphi(c \wedge d) > q$. If $\varphi(c \wedge d) = q$, and as $\varphi(c \wedge d) \geqq c \wedge d$, also $c \wedge d = q$, and so $c \wedge d \in M_{qp}$ according to its definition. If $\varphi(c \wedge d) > q$, then $\varphi(c \wedge d) = \varphi(c) = \varphi(d)$, as in the other case $\varphi(c) > \varphi(c \wedge d) > \lambda(c)$, where $\varphi(c \wedge d) \in R$, which is a contradiction. So $\lambda(c \wedge d) = q$ and $\varphi(c \wedge d) = p$, whence $c \wedge d \in M_{qp}$. The proof is analogous for $\lambda(c \vee d)$.

2° Let $c = p$, $\varphi(d) = p$ and $\lambda(d) = q$. As $\varphi(d) \geqq d$, $c \vee d = c$ and $c \wedge d = d$. Consequently, $\varphi(c \wedge d) = p$, $\lambda(c \wedge d) = q$, and $c \vee d = p$. So $c \wedge d$, $c \vee d \in M_{qp}$. The proofs for other combinations are similar.

Accordingly, if $c, d \in M_{qp}$, then $c \wedge d$, $c \vee d \in M_{qp}$. According to the definitions of φ, λ and M_{qp}, $q \leqq x \leqq p$ for any $x \in M_{qp}$. Let $q \leqq h \leqq p$ and $h \in L$. According to (ii) and (ii′),

$$q \leqq \lambda(h) \leqq h \leqq \varphi(h) \leqq p\,.$$

As $\lambda(h)$, $\varphi(h) \in R$, $q < \lambda(h) \leqq \varphi(h) < p$ cannot hold, and so $h \in \{q, p\}$, or $\varphi(h) = p$ and $\lambda(h) = q$. Therefore $h \in M_{qp}$. This completes the proof.

The family $\mathcal{M}$ determined by a screen R of L does not induce a homomorphism on L, but a *tolerance relation* as the following theorem shows. For tolerance relations on lattices and abstract algebras and for mappings generated by tolerance relations the reader is referred to the paper [2] of Chajda, Niederle and Zelinka and to the references of that paper. Numerous results on tolerance relations are also given in the paper [4] of Chajda and Zelinka. We shall show that $\mathcal{M}$ generates a binary relation T on L as follows: $aTb \Leftrightarrow a, b \in M_{qp}$ for some $M_{qp} \in \mathcal{M}$. T is a compatible tolerance relation (briefly, a tolerance relation) on L if it is reflexive, symmetric and compatible on L (i.e. aTb and cTd imply $a \vee cTb \vee d$ and $a \wedge cTb \wedge d$).

Theorem 3: *Let L be a lattice, R a screen of L, and φ and λ the mappings characterizing R. There are no elements $a, b, a \in L$ and $b \in R$, such that $\varphi(a) > b > \lambda(a)$ if and only if the family $\mathcal{M}$ generates a tolerance relation T on L, where $aTb \Leftrightarrow a, b \in M_{qp}$ for some $M_{qp} \in \mathcal{M}$.*

Proof: If $\mathcal{M}$ generates a tolerance relation on L, the tolerance classes $M_{qp} \in \mathcal{M}$ of T are convex sublattices of L [3, Thm. 1]. The desired result follows now from Lemma 1.

Let R be a screen of L such that $\varphi(a) > b > \lambda(a)$ holds for no $a \in L$ and no $b \in R$. The definitions of the sets M_{qp} and of the family $\mathcal{M}$ show that $\bigcup \{M_{qp} \mid M_{qp} \in \mathcal{M}\} = L$, and so T is reflexive. The symmetry of T is evident.

In the following we assume that $q \neq p$ and $r \neq t$ as the other cases are obvious. Let aTb and cTd, i.e. there are two sets M_{qp}, $M_{rt} \in \mathcal{M}$ such that $a, b \in M_{qp}$ and $c, d \in M_{rt}$. As M_{rt} and M_{qp} are convex sublattices of L, we can assume that $a \geqq b$ and $c \geqq d$. Suppose that there were an element $h \in R$ such that $a \vee c \geqq h \geqq b \vee d$. According to the definitions of M_{qp} and M_{rt}, $q < p$ and $r < t$ and so

$$\varphi(t \vee p) \geqq t \vee p \geqq a \vee c \geqq h \geqq b \vee d \geqq r \vee q \geqq \lambda(r \vee q) .$$

According to the properties of L, $h = \varphi(t \vee p)$, or $h = \lambda(r \vee q)$, or there exists no element $h \in R$ with the property $a \vee c \leqq h \leqq \leqq b \vee d$. In each case given by h, $a \vee c$ and $b \vee d$ belong to one and same set M of the family $\mathcal{M}$, whence $a \vee cTb \vee d$. The proof is similar for $a \wedge cTb \wedge d$. Thus T is compatible and also a tolerance relation on L. This completes the proof.

The following lemma illuminates the structure of the sets $M_{qp} \in \mathcal{M}$.

Lemma 2: *Let L be a lattice, R a screen of L generating a tolerance relation T on L by means of $\mathcal{M}$. If $M_{qp} \cap M_{rt} \neq \emptyset$ for some M_{qp}, $M_{rt} \in \mathcal{M}$ ($q \neq r$ and $p \neq t$), then $M_{qp} \cap M_{rt} \subset \{q, p, r, t\}$.*

Proof: Let $x \in M_{qp} \cap M_{rt}$, $q, r < x < p, t$. Then $p, t \in [x) \cap R$ and they are both least elements in this set, whence $p = t$. Similarly we see that $q = r$, which is a contradiction.

Now we can extend the result of [5, Thm. 7].

Theorem 4: *Let R be a screen of a lattice L, λ and φ the mappings characterizing R, and β an optimal rounding $L \to R$. $\beta(a) \in \{\lambda(a), \varphi(a)\}$ for every $a \in L$ if R determines a tolerance relation on L.*

Proof: As $\lambda(a) \leqq a \leqq \varphi(a)$ for every $a \in L$ and R is the set of fixpoints of β,

$$\lambda(a) = \beta(\lambda(a)) \leqq \beta(a) \leqq \beta(\varphi(a)) = \varphi(a) .$$

According to Theorem 3 there are no $a \in L$ and no $b \in R$ such that $\lambda(a) < b < \varphi(a)$, and as β is a mapping $L \to R$,

$$\beta(a) \in \{\lambda(a), \varphi(a)\} .$$

As one can easily see, the rounding β of Theorem 4 divides each $M_{qp} \in \mathcal{M}$ into two disjoint subsets M_q and M_p as follows (cf. [5, Thm. 7]): $M_q \cup M_p = M_{qp}$, $\beta(h) = q = \lambda(h)$ for each $h \in M_q$ and $\beta(h) = p = \varphi(h)$ for each $h \in M_p$. Further, $c \leqq d$ holds for no pair c, d, $c \in M_p$ and $d \in M_q$.

Let us consider the lattice L of Figure 1, where $R = \{x, y, z, w\}$ is a screen. R does not determine a tolerance relation on L, but as $d < w, a$, $\beta(h) \in \{\lambda(h), \varphi(h)\}$ for every $h \in L$ and every optimal rounding $\beta: L \to R$. Thus the condition of Theorem 4 is not necessary.

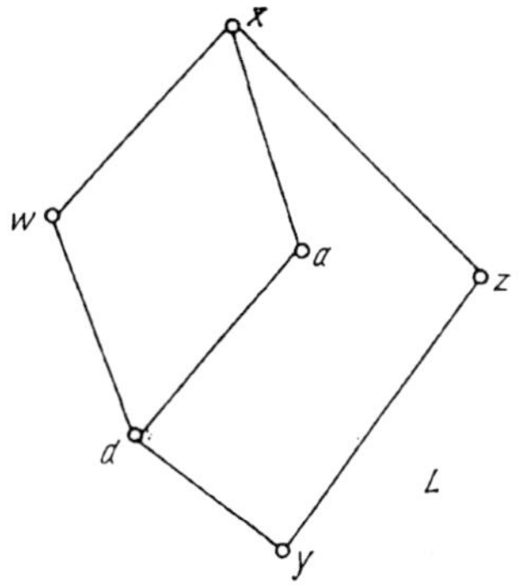

Fig. 1

References

1 Apostolatos, N., Einige Ergänzungen zu dem Begriff des Rasters, ZAMM 54, 517–518 (1974).
2 Chajda, I., Niederle, J.; Zelinka, B., On existence conditions for compatible tolerances, Czech. Math. J. 26, 304–311 (1976).
3 Chajda, I.; Zelinka, B., Tolerance relations on lattices, Časopis pést. mat. 99, 394–399 (1974).
4 —, Lattices of tolerances, Časopis pést. mat., 102, 10–24 (1977).
5 Kulisch, U., An axiomatic approach to rounded computations, Numer. Math. 18, 1–17 (1971).
6 —, On the concept of a screen, ZAMM 53, 115–119 (1973).

Eingereicht: 16. 2. 1977

Anschrift: Dr. J. Nieminen, Vaasa school of economics, Institute of mathematics, Raastuvank. 31, 65100 Vaasa 10, Finland

ZAMM 58, 520–522 (1978)

P. S. Theocaris / N. I. Ioakimidis

Application of the Gauß, Radau and Lobatto Numerical Integration Rules to the Solution of Singular Integral Equations

1. Introduction-Statement of the problem

Cauchy-type singular integral equations are often encountered in several branches of Mathematical Physics. Nevertheless, their numerical solution presents considerable difficulties and very few efficient methods have been proposed. An account of most of the existing methods for the numerical solution of such equations can be found in a monograph by Ivanov [1]. Unfortunately, these methods are almost inapplicable to Cauchy-type singular integral equations along a real interval. For this class of integral equations the method of reduction to an equivalent Fredholm integral equation, in accordance with the developments of Muskhelishvili [2], can be used. An effective algorithm for the numerical solution of Cauchy type singular integral equations, based on the theoretical investigations of Muskhelishvili and the reduction of the singular integral equation to a Fredholm integral equation, was recently developed by Dow and Elliott [3, 4].

On the other hand, some direct methods for the numerical solution of Cauchy-type singular integral equations along a real interval have been proposed. These methods consist in approximating such an equation by a system of linear equations, either by applying the integral equation at a number of collocation points, or by expanding the unknown function into a series of orthogonal polynomials [5, 11]. An account of these methods can be found in the paper by Erdogan, Gupta and Cook [5].

Recently, Majumdar [6] proposed a method for the numerical solution of a class of Cauchy-type singular integral equations, namely equations of the form:

$$\int_{-1}^{1} K(x - t)\, \varphi(t)\, dt = f(x), \qquad -1 < x < 1 , \tag{1}$$

where the kernel $K(x - t)$ consists of a singular part $a/(x - t)$

plus a regular part $K^*(x-t)$, that is:

$$K(x-t) = \frac{a}{x-t} + K^*(x-t)\,. \tag{2}$$

In equation (1) the unknown function $\varphi(t)$ is assumed to be regular along the interval $[-1,1]$ and the right-hand side function $f(x)$ is also assumed to be regular along $(-1,1)$ besides in the neighbourhoods of the end-points $x = \pm 1$, where it may present logarithmic singularities.

For the numerical solution of equation (1) MAJUMDAR [6] proposed its reduction to a system of linear equations after an expansion of the functions $K^*(x-t)$ and $\varphi(t)$ into polynomial series. Then $f(x)$ can also be expanded into a series form. Next the integral equation is applied at a number of arbitrarily selected collocation points and, in this way, it is reduced to a system of linear equations. This technique presents the following disadvantages:

i) The kernel $K(x-t)$ is a function of the difference $(x-t)$ and not of the variables x and t independently as is usually the case.

ii) The computation of the coefficients in the expansion of $K^*(x-t)$ into a polynomial series is a difficult task but not necessary as will be seen later.

iii) The evaluation of a system of special functions introduced by the author, several times, considerably increases the computational effort required.

All these disadvantages of the method proposed by MAJUMDAR for the numerical solution of equation (1), under the above-mentioned conditions, disappear if the numerical solution of this equation is based on its reduction to a system of linear equations by using a GAUSSian numerical integration rule for the approximation of the integral in (1), accompanied by a proper selection of the collocation points used. It may also be noted that the method proposed for the numerical solution of (1) is not a special case of the methods proposed in [5], which do not apply to the class of singular integral equations considered by MAJUMDAR [6].

2. The method of solution

Consider the well-known GAUSS numerical integration rule for regular integrals:

$$\int_{-1}^{1} \varphi(t)\,\mathrm{d}t \cong \sum_{i=1}^{n} A_i\varphi(t_i)\,, \tag{3}$$

where the abscissae t_i and the weights A_i, have been completely tabulated for several values of the number n of abscissae used [7]. By taking into account the developments of HUNTER [8], who generalized the GAUSS numerical integration rule to the case of CAUCHY-type principal value integrals, it is easy to see that equation (3) remains valid even if its left-hand side is a CAUCHY-type principal value integral of the form:

$$\int_{-1}^{1} \varphi(t)/(t-x)\,\mathrm{d}t\,,$$

provided that x is not arbitrary but it is selected as a root of the LEGENDRE function of the second kind $Q_n(x)$. The properties of this function can be found in [9]. In the same way it can be proved that the RADAU and LOBATTO numerical integration rules, the abscissae and weights of which have been also extensively tabulated in [7], can also be used for the evaluation of CAUCHY-type principal value integrals, provided that x is properly selected. It can be shown that x should be selected as a root of the function $[Q_n(x) + Q_{n-1}(x)]$ for the RADAU rule with an abscissa coinciding with $t = -1$, as a root of the function $[Q_n(x) - Q_{n-1}(x)]$ for the same rule but with an abscissa coinciding with $t = 1$, and as a root of the function $[Q_n(x) - Q_{n-2}(x)]$ or the function $\frac{\mathrm{d}}{\mathrm{d}x} Q_{n-1}(x)$ in the case of the LOBATTO rule.

In accordance with a theorem reported by SZEGÖ [9], the LEGENDRE function of the second kind and of degree n, $Q_n(x)$, has $(n+1)$ roots inside the interval $(-1,1)$, alternating with the roots of the corresponding LEGENDRE polynomial $P_n(x)$, which are the abscissae t_i used in the GAUSS numerical integration rule [7]. Furthermore, by taking into account well-known properties of the LEGENDRE functions [9], we can show that in the case of the RADAU rule there exist n roots of the function $[Q_n(x) \pm Q_{n-1}(x)]$, lying inside the interval $(-1,1)$ and alternating with the abscissae used. Finally, in the case of the LOBATTO rule there exist $(n-1)$ roots of the function $[Q_n(x) - Q_{n-2}(x)]$, lying inside the interval $(-1,1)$ and alternating with the corresponding abscissae, too.

In accordance with these developments, we can use the classical algebraic method of solution of FREDHOLM integral equations, that is the method of reduction of such an equation to a system of linear equations by using appropriate collocation points and approximating the integral through a numerical integration rule [10] for the solution of equation (1). Thus, we obtain:

$$\sum_{i=1}^{n} KA_i(t_i, x_k)\,\varphi(t_i) \cong f(x_k), \qquad k = 1, 2, \ldots, n\,, \tag{4}$$

where the kernel in equation (1) was considered as an arbitrary function $K(t, x)$ of the variables t and x, although this function, except its CAUCHY part, is assumed regular along the interval $[-1,1]$.

In (4) the collocation points x_k are selected in accordance with the foregoing developments. In most cases it seems preferable to use one of the two forms of the RADAU rule for the approximation of the integral in (1), since in this case there exist just n collocation points x_k. If one makes use of the GAUSS rule, one of the $(n+1)$ collocation points should be ignored. Finally, the LOBATTO rule, providing only $(n-1)$ collocation points x_k, can be used only when one more linear equation is available, most probably from a physical condition.

After the numerical solution of the system of equations (4) and the determination of the unknown function $\varphi(t)$ at the abscissae t_i of the numerical integration rule used, it is possible to find an expression for $\varphi(t)$, valid along the whole interval $[-1,1]$, through well-known interpolation techniques. Unfortunately, it seems very difficult to prove the convergence of the above-described numerical method to the correct results.

3. Two applications

We consider the following singular integral equations, already considered by MAJUMDAR [6]:

$$\int_{-1}^{1} K(x-t)\,\varphi(t)\,\mathrm{d}t = \frac{\pi x}{2} + \frac{\pi x - 2\ln\left|\frac{1-x}{1+x}\right|}{2(1+x^2)}\,, \quad -1 < x < 1\,, \tag{5}$$

$$\int_{-1}^{1} K(x-t)\,\varphi(t)\,\mathrm{d}t = (x^2 + 2x - 1)\ln\left|\frac{1+x}{1-x}\right| - \frac{10}{3}x - \frac{16}{3}\,, \quad -1 < x < 1\,, \tag{6}$$

$$\text{with } K(x-t) = \frac{1}{x-t} + x - t\,. \tag{7}$$

These equations have the exact solutions [6]:

$\varphi(t) = 1/(1+t^2)$ and $\varphi(t) = t^2 + 2t - 1$, (8), (9)

respectively.

The RADAU numerical integration rule with $n = 10$ abscissae, one of which coincided with the point $t = -1$, was used for the numerical solution of equations (5) and (6). The numerical results for equation (6), having as solution a polynomial of the third degree, (cf. (9)), have been seen to be completely accurate as expected. As regards equation (5), the numerical results for the unknown function $\varphi(t)$ at the abscissae used differed from the exact values of $\varphi(t)$ by no more than 0.25×10^{-6}, while the corresponding error for the same number of linear equations in the results given by MAJUMDAR was 0.6. After the approximation of $\varphi(t)$ along the whole interval $[-1, 1]$ through an interpolation polynomial, the maximum error was increased to 1.11×10^{-3} but, evidently, it remained much less than the error resulted when using the method of MAJUMDAR.

References

1 IVANOV, V. V., The Theory of Approximate Methods and Their Application to the Numerical Solution of Singular Integral Equations, Noordhoff, Leyden 1976 Translation of the Russian edition: Naukova Dumka, Kiev 1968.

2 MUSKHELISHVILI, N. I., Singular Integral Equations, Noordhoff, Leyden 1977 Translation of the Russian edition: Fizmatgiz, Moscow 1962.

3 DOW, M. L.; ELLIOTT, D., The numerical solution of singular integral equations over (−1, 1), Technical report No. 91, The University of Tasmania, Hobart 1977.

4 DOW, M. L., Singular integral equations, Ph. D. thesis, The University of Tasmania, Hobart 1977.

5 ERDOGAN, F.; GUPTA, G. D.; COOK, T. S., Numerical solution of singular integral equations, in: Methods of Analysis and Solutions of Crack Problems (Vol. 1 in the series: Mechanics of fracture, edited by SIH, G. C.), Noordhoff, Leyden 1973.
6 MAJUMDAR, S. R., Note on the numerical solution of singular FREDHOLM linear integral equation, ZAMM 54, 515–517 (1974).
7 STROUD, A. H.; DON SECREST, GAUSSIAN Quadrature Formulas, Prentice Hall, Englewood Cliffs (New Jersey) 1966.
8 HUNTER, D. B., Some GAUSS-type formulae for the evaluation of CAUCHY principal values of integrals, Numer. Math. 19, 419–424 (1972).
9 SZEGÖ, G., Orthogonal Polynomials (2nd and revised edition), American Mathematical Society, New York 1959.
10 KANTOROVICH, L. V.; KRYLOV, V. I., Approximate Methods of Higher Analysis, Noordhoff, Groningen 1958 Translation of the Russian edition: Fizmatgiz, Moscow-Leningrad 1962.
11 THEOCARIS, P. S., IOAKAMIDIS, N. I., Numerical integration methods for the solution of singular integral equations, Quart. Appl. Math. 35, 173–183 (1977).

Eingereicht: 5. 8. 1976, revidierte Fassung 30. 8. 1977

Anschriften: Prof. Dr. P. S. THEOCARIS; Dr. N. I. IOAKIMIDIS, Department of Theoretical and Applied Mechanics, The National Technical University of Athens, 5 K. Zographou Street, Zographou, Athens (625), Greece

ZAMM 58, 522 (1978)

TYN MYINT-U

An Alternative Method on Complete Integrals of Plane Eikonal Equations

1. Purpose

A class of complete integrals of the plane eikonal equation

$$|\text{grad } u|^2 = f^2(x, y) \tag{1}$$

for harmonic function $\ln f$ has been determined recently using the complex variables [1]. The purpose of this note is to apply an alternative method to the same equation without using the complex variables.

2. Analysis

Let $u(x, y) \in C^2$ in a domain D such that

$$u_x = f \cos \omega\,, \qquad u_y = f \sin \omega \tag{2), (3}$$

for some function $\omega(x, y) \in C^2$ in D. Since $u_{xy} = u_{yx}$, we have for $f(x, y) \in C^2$ in D the relation

$$(f \cos \omega)\, \omega_x + (f \sin \omega)\, \omega_y = f_y \cos \omega - f_x \sin \omega \tag{4}$$

which is a first order partial differential equation.

The characteristic equations for $f(x, y) \neq 0$ in D is given by

$$\frac{dx}{f \cos \omega} = \frac{dy}{f \sin \omega} = \frac{d\omega}{(f \cos \omega)\, \omega_x + (f \sin \omega)\, \omega_y}\,. \tag{5}$$

Consequently from equations (4) and (5) we obtain

$$\frac{dy}{dx} = \tan \omega\,, \qquad \frac{d\omega}{dx} = \frac{f_y}{f} - \frac{f_x}{f} \tan \omega\,. \tag{6), (7}$$

Differentiation of (6) using (7) yields

$$\frac{dy'}{1 + y'^2} = \frac{f}{f_y}\, dx - \frac{f_x}{f}\, dy\,. \tag{8}$$

This is an ordinary differential equation which can be integrated if the right side is an exact differential [2], [3]. That is, there is some function $U(x, y) \in C^2$ in D such that

$$U_x = \frac{f_y}{f} = \frac{\partial \ln f}{\partial y} = F_y\,, \tag{9}$$

$$U_y = -\frac{f_x}{f} = -\frac{\partial \ln f}{\partial x} = -F_x\,. \tag{10}$$

These equations are the CAUCHY-RIEMANN equations and hence U and F are harmonic.

Integration of equation (9) with respect to x and using equation (10) yields

$$U(x, y) = \int_{x_0}^{x} F_y(s, y)\, ds - \int_{y_0}^{y} F_x(x_0, t)\, dt + c_1\,, \tag{11}$$

where x_0, y_0 and c_1 are constants. Now integrating equation (8) we obtain

$$\arctan y' = U + c_2\,,$$

where c_2 is a constant. Since $y' = \tan \omega$, we have

$$\omega(x, y) = U(x, y) + \alpha \tag{12}$$

where α is a constant. With this relation and the CAUCHY-RIEMANN equations (9) and (10) it can easily be shown that $u(x, y)$ is harmonic.

Integrating equation (2) using equation (3) we finally obtain

$$u(x, y) = \int_{x_0}^{x} f(s, y) \cos (U(s, y) + \alpha)\, ds + \int_{y_0}^{y} f(x_0, t) \sin (U(x_0, t) + \alpha)\, dt + C\,, \tag{13}$$

where C is a constant.

3. Illustrations

(a) Consider the equation

$$u_x^2 + u_y^2 = e^{2kx}$$

with constant k. Here $F = \ln e^{kx} = kx$ and hence

$$U = -\int_{y_0}^{y} k\, dt = -ky + c_1\,.$$

Thus we have

$$u(x, y) = \int_{x_0}^{x} e^{ks} \cos (-ky + \alpha)\, ds + \int_{y_0}^{y} e^{kx_0} \sin (-kt + \alpha)\, dt + C$$
$$= \frac{a}{k}\, e^{kx} \cos ky \pm \frac{\sqrt{1 - a^2}}{k}\, e^{kx} \sin ky + b\,,$$

where a and b are constants.

(b) Next consider the equation

$$u_x^2 + u_y^2 = x^2 + y^2\,, \qquad r = \sqrt{x^2 + y^2}\,.$$

In this case $F = \ln r$ and U is given by

$$U(x, y) = \int_{x_0}^{x} \frac{y}{s^2 + y^2}\, ds - \int_{y_0}^{y} \frac{x_0}{x_0^2 + t^2}\, dt + c_1 = \arctan (x/y) + c_2,$$

where c_2 is a constant. Thus we have

$$u(x, y) = \int_{x_0}^{x} \sqrt{s^2 + y^2} \cos \left(\arctan \left(\frac{s}{y}\right) + \alpha\right) ds + \int_{y_0}^{y} \sqrt{x_0^2 + t^2} \sin \left(\arctan \left(\frac{x_0}{t}\right) + \alpha\right) dt + C$$
$$= \frac{a}{2}\, (x^2 - y^2) \pm \sqrt{1 - a^2}\, xy + b$$

where a and b are constants.

4. Conclusion

A class of complete integrals of plane eikonal equations for harmonic functions $\ln f$ are determined without using the complex variables. Two examples are presented. The integrations encountered here are found to be more difficult than that of the complex integrals in [1]. However, the analysis here is simple and direct.

References

1 MYINT-U, T., A class of complete integrals of plane eikonal equations, Internat. J. Nonlinear Mech. 12, 33–36 (1977).
2 GHAFFARI, A., On the integrability cases of the equation of motion for a satellite in an axially symmetric gravitational field, Celest. Mech. 4, 49–53 (1971).
3 SERRET, J., Lehrbuch der Differential- und Integralrechnung, Vol. 3, Leipzig 1909.

Eingereicht: 23. 5. 1977

Anschrift: Dr. T. MYINT-U, Manhattan College, Department of Mathematics and Computer Science, Riverdale, New York 10471, U.S.A.

BUCHBESPRECHUNGEN

Szabó, I., Geschichte der mechanischen Prinzipien und ihrer wichtigsten Anwendungen. Basel-Stuttgart. Birkhäuser Verlag. 1977. XV, 491 S., Fr. 48,–.

Dieses schöne und inhaltsreiche Werk des bekannten Verfassers enthält wesentlich mehr als der Titel vermuten läßt, nämlich außer der Geschichte der Prinzipien der Mechanik noch eine ausführliche Darstellung der Entwicklung der Hydromechanik und der Elastizitätstheorie.

Im ersten Kapitel behandelt der Autor die Grundlegung der klassischen Mechanik des starren Körpers durch Newton, Euler und d'Alembert. Er hebt besonders die wichtige Rolle L. Eulers hervor, auf den so fundamentale Sätze wie das Schnittprinzip und der Momentensatz zurückgehen; hierauf hat wohl zuerst C. A. Truesdell hingewiesen. Zur einwandfreien Ableitung dieses Satzes für ein festes Kontinuum benötigt man die Symmetrie des Spannungstensors. Diese ergibt sich in der Statik aus den Gleichgewichtsbedingungen; in der Kinetik muß man sie nach L. Boltzmann als Axiom einführen. Eine solche einwandfreie Ableitung des Momentensatzes gibt z. B. G. Hamel in seinem immer noch lesenswerten Buch „Elementare Mechanik".

Ausführlich geht der Verfasser auf die Schwierigkeiten ein, die mit dem Begriff „Massenpunkt" verbunden sind, und empfiehlt, das zweite Newtonsche Grundgesetz von vornherein für die Translationsbewegung des Massenmittelpunktes auszusprechen.

Im letzten Abschnitt des ersten Kapitels geht der Verfasser auf das Prinzip von d'Alembert ein und betont ausdrücklich die wichtige Rolle, die Jacob und Daniel Bernoulli bei der Entstehung dieses Prinzipes gespielt haben. In Jacob Bernoullis Ermittlung des Schwingungsmittelpunktes eines Körperpendels und in Daniel Bernoullis Untersuchungen der Schwingungen einer Kette kommen schon Betrachtungen vor, die bei Verallgemeinerung zum d'Alembertschen Prinzip führen. Schließlich wird noch Lagranges Kombination des d'Alembertschen mit dem der virtuellen Verschiebungen besprochen.

Das zweite Kapitel beginnt mit einer Schilderung der nicht immer der Entwicklung förderlichen Rolle, die die Philosophen in der Mechanik gespielt haben. Z. B. widersprechen manche Sätze von Descartes in seiner Stoßtheorie der Erfahrung. Es folgt die Besprechung des Streites der Cartesianer und der Leibnizianer um das wahre „Kraftmaß", dessen Ursache darin lag, daß Begriffe wie „Kraft", „Bewegungsgröße bzw. Impuls" und „lebendige Kraft (besser kinetische Energie)" noch nicht klar definiert waren. Erst Daniel Bernoulli klärte diesen Streit durch den Hinweis auf den Unterschied zwischen Zeit- und Wegintegral der Kraft. Heute ist das fast trivial, damals war es dies aber nicht. Befaßt man sich mit der historischen Entwicklung der Grundbegriffe der Mechanik und den hierbei aufgetretenen Schwierigkeiten, so wundert man sich weniger darüber, daß die Mechanik bei den Studenten der Ingenieurwissenschaften das gefürchtetste Fach ist und in ihm die meisten „Durchfälle" auftreten.

In den folgenden Abschnitten des zweiten Kapitels werden zunächst die Geschichte des Prinzips von Maupertuis und der Prioritätenstreit zwischen ihm und J. S. Koenig behandelt. Dies leitet über zu den Variationsprinzipien von Hamilton, Gauss und Hertz. Die Hamilton-Jacobische Theorie, die bei Begründung der Quantenmechanik eine wichtige Rolle gespielt hat, wird nicht mehr besprochen, da sonst das Werk wohl zu umfangreich geworden wäre.

Das dritte, der Hydromechanik gewidmete Kapitel, beginnt mit dem Bericht über die Entwicklung der Hydrostatik durch Archimedes, Stevin und B. Pascal, wobei der Verfasser auch auf die schwierigen Stabilitätsuntersuchungen an schwimmenden Körpern von Archimedes hinweist, wobei dieser endliche Verschiebungen benutzte.

Den Übergang zur vollständigen Hydromechanik bilden Newtons Untersuchungen in den „Prinzipia"; dort findet man u. a. das quadratische Widerstandsgesetz und den Ansatz für die Reibung. Daß in ihnen auch falsche Ergebnisse enthalten sind, ist in einem Pionierwerk wie den „Prinzipia" nicht anders zu erwarten.

Der Energiesatz in Form der Bernoullischen Gleichung spielt in der Hydromechanik eine wichtige Rolle. Meist wird sie Daniel Bernoulli zugeschrieben; aus der „Hydraulica" von Johann Bernoulli, dem Vater von Daniel, geht aber nach Szabó hervor, daß man sie vielleicht besser nach diesem benennt. Insbesondere finden sich bei Johann zum erstenmal Betrachtungen am Flüssigkeitselement, die dann Euler zur Aufstellung der Grundgleichungen der Hydromechanik inspirierten. Andererseits finden sich bei Daniel die ersten Untersuchungen über elastische Flüssigkeiten und Ansätze zur kinetischen Gastheorie.

Nach Besprechungen der ersten Untersuchungen zur äußeren Ballistik wendet sich der Verfasser zur Vollendung der klassischen Hydrodynamik durch Euler und d'Alembert, der zum ersten Mal Funktionen von komplexem Argument auf ebene Strömungsprobleme anwendete; das D-Alembertsche Paradoxon findet sich allerdings schon in Anmerkungen Eulers zu einem Werk über Ballistik von Robins.

Der Aufstellung der Theorie idealer Flüssigkeiten folgte im 19. Jahrhundert die Theorie der zähen Flüssigkeiten, die von Navier, Saint-Venant und Stokes aufgestellt wurde. Während Navier und Stokes noch molekulartheoretische Vorstellungen benutzten, ging Saint-Venant von den Grundbegriffen der Kontinuumsmechanik aus und beschritt also den heute noch üblichen Weg. Nach Szabó sollte man also eigentlich von den Navier-St.-Venantschen Gleichungen sprechen. Der Lösung dieses nichtlinearen Systems partieller Differentialgleichungen stellen sich außerordentliche Schwierigkeiten entgegen, die erst durch die genialen Gedanken von L. Prandtl wenigstens z. T. überwunden wurden. Auf diese und andere Entwicklungen wird hingewiesen.

Bei dem nichtlinearen Problem der Gasdynamik brachte B. Riemanns berühmte Arbeit nach Untersuchungen von Earnshow wenigstens im eindimensionalen Fall den ersten großen Fortschritt. Auf diese Arbeit, durch die die Verdichtungsstöße erklärt wurden, geht der Verfasser ausführlich ein, ebenso auf die Arbeiten von E. Mach, der bei seinen sorgfältigen Experimenten die Schlierenmethode von Toepler verwendete. Schließlich wird noch auf die wichtigen Veröffentlichungen von J. Hadamard und R. Becker zur Gasdynamik und speziell den Verdichtungsstößen hingewiesen.

Bei der Entwicklung der Mechanik fester Continua spielte die schwingende Saite eine wesentliche Rolle. Schon Wissenschaftler des 17. Jahrhunderts hatten richtige Vorstellungen über die auftretenden Erscheinungen. Der Verfasser berichtet ausführlich über die Entwicklung der Theorie durch die großen Forscher des 18. Jahrhunderts, die u. a. schließlich zur Entwicklung einer Funktion nach Orthogonalfunktionen mit der Fourier-Reihe als erstem Beispiel führte.

Wohl noch wichtiger als die Theorie der schwingenden Saite waren für die Entwicklung der Mechanik des festen verformbaren Körpers die Untersuchungen über den auf Biegung beanspruchten Balken. Der Autor berichtet ausführlich über diese Entwicklung, die mit Galilei begann und u. a. über Jacob Bernoulli, Euler und Coulomb zur Biegetheorie von Navier führte.

Wesentliche Fortschritte in der Theorie des verformbaren Festkörpers waren erst möglich, nachdem Cauchy den Tensorcharakter des Spannungszustandes erkannt und die Grundgleichungen der Elastizitätstheorie aufgestellt hatte. Hierüber wird eingehend berichtet, ebenso über das Entstehen der Plattentheorie. Für deren Entwicklung waren die Versuche von Chladni über die Klangfiguren ebener Platten von Bedeutung, durch die Sophie Germain zu ihren theoretischen Untersuchungen angeregt wurde. Kirchhoff gab dann die heute noch meist benutzte endgültige Form der Theorie, deren Widersprüche durch die allerdings kompliziertere Theorie von E. Reissner beseitigt wurden. Mit einem Ausblick auf die Arbeiten von Saint-Venant und Clebsch schließt der Verfasser den Bericht über die Entwicklung der Elastizitätstheorie ab und wendet sich zum letzten Kapitel seines Werkes, der Stoßtheorie. Da so fundamentale Begriffe wie Impuls und kinetische Energie erst im Verlauf vieler Jahrzehnte geklärt wurden, ist es nicht verwunderlich, daß mit der Erfahrung übereinstimmende Ergebnisse erst nach und nach gefunden wurden. Bemerkenswert ist, daß der wenig bekannte Gelehrte Marcus Marci de Kronland aus Prag schon z. T. richtige Schlüsse aus Versuchen an Billardkugeln zog. Dagegen sind die sog. „Stoßgesetze" von Descartes entweder falsch oder inkorrekt. Erst die Untersuchungen von Wallis, Wren, Huygens und Euler brachten die richtige (phänomenologische) Theorie des elastischen Stoßes.

Die phänomenologische Stoßtheorie gibt keine Auskunft über die Stoßdauer, die Stoßkraft und den Spannungs- und Verformungszustand in den stoßenden Körpern. Die ersten Vorstöße in dieses schwierige Gebiet gelangen den Physikern F. Neumann und H. Hertz, deren Ergebnisse innerhalb der jeweiligen Gültigkeitsgrenzen auch durch Versuche bestätigt wurden. Mit einem Ausblick auf die weitere Entwicklung der Stoßtheorie verformbarer Körper, deren experimentelle Untersuchung erst durch die

Entwicklung der Elektronik möglich wurde, schließt das ausgezeichnete Werk des Verfassers. Sein Studium sei jedem empfohlen, der sich für die Geschichte der Mechanik interessiert. Sehr lobenswert ist die hervorragende Ausstattung des Buches durch den Verlag.

Eichwalde A. WEIGAND

Schäfer, W., Theoretische Grundlagen der Stabilität technischer Systeme. Berlin. Akademie-Verlag. 1976. 130 S., M 8,–. (WTB Mathematik–Physik 169).

Das als Nachschlagewerk für Ingenieure, Verfahrenstechniker, Chemiker u. a. konzipierte Taschenbuch bringt eine Zusammenstellung von Definitionen und Sätzen, die für die Anwendung der sogenannten direkten Methode von LJAPUNOW zur Untersuchung der Stabilität von Systemen relevant sind.

Im ersten Kapitel werden mathematische Grundlagen zitiert und wesentliche Begriffe wie Bewegungen, dynamische und allgemeine Systeme sowie ihre Beschreibung durch Differential- und Differenzengleichungen eingeführt. Im zweiten folgen die Stabilitätsbegriffe für die zuvor unterschiedenen Bewegungen und Systeme. Kapitel 3 erläutert den Grundgedanken der Methode von LJAPUNOW und die Konstruktion von LJAPUNOW-Funktionen am Beispiel linearer Differential- und Differenzengleichungen. Die Kapitel 4 bis 9 enthalten die Stabilitätsbedingungen für die einzelnen mathematischen Modelle, wobei auch — bei gewöhnlichen Differentialgleichungen — kurz auf die Existenz von LJAPUNOW-Funktionen und die Bestimmung von Einzugsgebieten eingegangen wird.

Der Autor hat sorgfältig Begriffsbildungen zur Stabilität zusammengetragen und ihnen Stabilitätsbedingungen zugeordnet. Im Hinblick auf den angesprochenen Leserkreis (Beweise werden nicht geführt) nimmt die Darstellung der Leistungsfähigkeit der behandelten Methode nur geringen Raum ein. Die Ausführung des im Titel angesprochenen technischen Bezuges wäre hier vom Leser dankbar aufgenommen worden.

Berlin G. LIPPOLD

Oswatitsch, K., Grundlagen der Gasdynamik, Wien-New York. Springer-Verlag. 1976. XIII, 725 S., 309 Abb., S 2040.–. DM 296,–.

This book is a new, modernized, enlarged and improved edition of the very successful, translated into many languages, „Gasdynamik" of the same author which appeared a quarter of a century ago and was distinguished by an introduction of one of the Founders of this field of mechanics — Ludwig PRANDTL. It does not cover the whole field as the first book tried to do, in particular transsonic flows, three-dimensional flows as well as experimental methods are left out entirely. As mentioned in the author's introduction, transsonic flows and hypersonic flows will be included in an accompanying volume „Spezialgebiete der Gasdynamik". Although the titles of the remaining chapters remain the same, they are greatly enlarged, nearly doubled, and new developments are considered. The first chapter on thermodynamics is nearly quadrupled and gives a good background to the analysis of different relaxation processes and radiation processes connected with gas flows.

The governing equations are deduced gradually for the simpler problems in the first parts of the book to their more general forms later on. The integral method of approach for the deduction of the basic relations is preferred and differential divergence type equations are often used. About one third of the book is devoted to stationary one-dimensional flows including relaxation effects and radiation. Great attention is paid to the plane, cylindrical and spherical one-dimensional shock wave problems without and with dissociation, ionisation, radiation effects, and examples of application in inner ballistics and engine design are given.

Another third of the book covers subsonic and supersonic plane and axisymmetric flows where wide use is made of perturbation methods and the method of characteristics.

The author uses selected references and preferred methods of solution without emphasis on mathematical aspects and generalisations. Undoubtedly the book fulfills the author's desire and gives a very useful treaty of the physical and technical problems of gasdynamics based on personal experience and teinted by the author's charming personality.

I would like to recommend this book strongly to all specialists in this field. Helas the extremely high price of the book, compensated partly by the high editorial and printing standard achieved, will not make it accessible, even in libraries, to all interested unless a more reasonably priced edition or translation appears.

Warszawa W. FISZDON

Findley, W. N. / Lai, J. S. / Onaran, K., Creep and relaxation of nonlinear viscoelastic materials with an introduction to linear viscoelasticity. Amsterdam-New York-Oxford. North-Holland. 1976. XIII, 367 S., Dfl. 85.–. US $ 34.50 (Applied Mathematics and Mechanics 18).

Die Verfasser geben eine systematische und übersichtliche Darstellung der Kriech- und Relaxationserscheinungen in viskoelastischen Materialien, wobei schwerpunktmäßig die Möglichkeiten zur analytischen Beschreibung der zeitabhängigen Materialeigenschaften vermittelt werden, aber auch auf die Lösung von Randwertproblemen und auf experimentelle Methoden eingegangen wird. Die Verfasser wenden sich sowohl an Wissenschaftler und Konstrukteure als auch an Studenten, was wohl ein Grund dafür ist, daß die Einführung in die lineare Theorie umfangreich ist.

Das Buch beginnt mit einer Erläuterung der viskoelastischen Grundbegriffe und einem Einblick in die Entwicklung der Viskoelastizitätstheorie bei Plasten und Metallen. Ausführlich wird auf die linearen viskoelastischen Konstitutivgleichungen eingegangen. Verwendung finden differentielle und integrale lineare Spannungs-Dehnungs-Beziehungen, die durch Kombination der Elemente Feder und Dämpfer und unter Benutzung des BOLTZMANNschen Überlagerungsprinzips erhalten werden. Eine Rolle spielen dabei auch das Verhalten bei dynamischer Belastung, komplexe Moduln, Energiedissipation und die Erfassung von Temperatureffekten. Einige Anwendungsbeispiele beschließen die Ausführungen zur linearen Viskoelastizitätstheorie. Die Lösungen werden mit Hilfe der LAPLACE-Transformation und des Korrespondenzprinzips ermittelt.

Zur Beschreibung nichtlinearer viskoelastischer Materialeigenschaften beschränken sich die Verfasser auf die Verwendung von Mehrfachintegralen. Auf diese Weise gelingt ihnen eine relativ abgeschlossene Darstellung des Hauptgegenstandes ihres Buches. Eingeführt wird das nichtlineare Kriechen bei konstanter Spannung bzw. die nichtlineare Relaxation bei konstanter Dehnung. Darauf aufbauend werden das nichtlineare Kriechen (bzw. Relaxation) bei veränderlicher Spannung (bzw. Dehnung) und das kombinierte Problem behandelt. Praktische Methoden zur Bestimmung der Kerne der Mehrfachintegrale werden erläutert. Zahlreiche experimentelle Ergebnisse, die in das Buch aufgenommen werden, untermauern die Bedeutung dieser Darstellung. Es werden Lösungsmöglichkeiten für das wichtige Problem veränderlicher Temperaturen bei nichtlinearen viskoelastischen Materialien vorgestellt. Auch die Ausführungen zur nichtlinearen Viskoelastizitätstheorie enden mit einem kurzen Kapitel zur Lösung von Randwertproblemen.

Das Buch schließt mit einigen Bemerkungen über experimentelle Methoden, wobei insbesondere Versuchsanlagen und Maschinen vorgestellt werden. Der Anhang, in welchem die benötigten mathematischen Grundlagen zusammengestellt werden, erleichtert die Einarbeitung in die behandelte Thematik.

Das Buch füllt zweifellos eine Lücke in der Literatur zur Viskoelastizitätstheorie. Es ist so abgefaßt, daß es dem Anliegen, einen breiten Leserkreis anzusprechen, gerecht wird. Geprüft werden sollte jedoch, ob in der nächsten Auflage die Kapitel über die Behandlung von Randwertproblemen durch weitere Anwendungsbeispiele ergänzt werden können.

Berlin W. KAISER

Onicescu, O./Guiaşu, S., Mecanique Statistique, Wien-New York. Springer-Verlag. 1971. VIII, 346 S., DM 71,50. $ 493,–. (CISM Monographs 1)

Die statistische Mechanik hat etwa in den letzten drei Jahrzehnten eine unerhörte Entwicklung sowohl in den Grundlagen wie in den vielfältigsten Anwendungen innerhalb aller Naturwissenschaften genommen. Das gilt auch für die mathematische Fundierung der Theorie mit den Mitteln der Funktionalanalysis. Das vorliegende Werk, als Monographie Nr. 1 des International Centre for Mechanical Sciences in Udine (Italien) mit dem Untertitel „Mathematische Prinzipien" erschienen, beschäftigt sich nicht mit der modernen Entwicklung der statistischen Mechanik innerhalb der mathematischen

Physik, sondern gibt in lehrbuchartiger Form eine Einführung in den klassischen Apparat der statistischen Mechanik, wobei sowohl die Gleichgewichtsstatistik (nach dem Vorbild des sowjetischen Wahrscheinlichkeitstheoretikers A. I. KHINTCHINE) wie auch die Statistik irreversibler Prozesse (mit BOLTZMANN-Gleichung, MARKOFF-Prozessen und FOKKER-PLANCK-Gleichung) behandelt werden. Probleme der Quantenstatistik finden nur einige Male am Rande Erwähnung.

Von den neueren Entwicklungen wird der informationstheoretische Zugang zur Gleichgewichtsstatistik abgehandelt; eine Erörterung der viel wichtigeren informationstheoretischen Methodik in der Statistik irreversibler Prozesse (ROBERTSON, SUBAREW) fehlt. Leider vermißt man auch modernere Konzeptionen wie die der thermodynamischen Greenschen Funktionen und Methoden wie die Projektionsoperatorformalismen nach ZWANZIG, MORI oder PRIGOGINE sowie die störungstheoretischen Techniken (einschließlich Graphentechniken), die heute alle eine beherrschende Rolle spielen.

Das Buch wird so seinem Untertitel nicht gerecht. Andererseits ist es wegen seiner meist recht gründlichen Diskussion schwieriger Grundlagenprobleme als Einführung und Ergänzungslektüre für alle, die der französischen Sprache mächtig sind, zu empfehlen.

Dresden G. VOJTA

Roseau, M., Solutions Periodiques ou Presque Periodiques des Systemes Differentiels de la Mecanique Non-Lineaire. Lectures held at the Dptt. for Mechanics of Deformable Bodies 1970, Wien-New York. Springer-Verlag. 1970. 228 S., DM 39,90. $ 275,—. (CISM Courses and Lectures 44)

Das Büchlein trägt den Charakter von lecture notes und ist aus Vorlesungen entstanden, die Verf. 1970 in Udine über das Thema periodische und fastperiodische Lösungen von nichtlinearen gewöhnlichen Differentialgleichungen gehalten hat. Genauer gesagt werden die drei wichtigsten analytischen Verfahren zum Nachweis der Existenz und gegebenenfalls zur Approximation solcher Lösungen vorgestellt. Die Darstellung ist knapp, aber lückenlos und verlangt vom Leser an Vorkenntnissen nicht mehr als das was zum Handwerkzeug der klassischen Analysis gehört. Insbesondere wird in den ersten beiden Kapiteln die Theorie des Abbildungsgrades nach dem Vorgang von E. HEINZ von Grund aus entwickelt. Dieser Teil des Buches ist auch dann lesenswert, wenn man nicht primär an speziellen analytischen Techniken bei Differentialgleichungen interessiert ist. Die Anwendungsbeispiele zur Abbildungsgradtheorie in den Abschnitten 5.1—5.3 sind recht instruktiv und ohne großen Aufwand zugänglich.

Bei dem ersten der vorgestellten Verfahren (Kap. 4 und 5) wird die Lösbarkeit der periodischen Randwertaufgabe auf die Lösbarkeit eines endlichen Gleichungssystemes (determinierende Gleichungen, Verzweigungsgleichungen) zurückgeführt. Es ließe sich dieses Verfahren daher auch in den größeren Zusammenhang der Randwertprobleme bei nichtlinearen Differentialgleichungen einordnen, und von dorther gesehen existieren Alternativen zu der vom Verf. ausschließlich behandelten Methode von CESARI. Es gibt jedoch gute Gründe, dieser Methode den Vorzug zu geben. Sie ist eine naturgemäße Weiterentwicklung von klassischen Ansätzen (GALERKIN-Verfahren, Störungsrechnung) und ist an konkreten und auf anderen Wegen nicht oder jedenfalls nicht so direkt angreifbaren Aufgabenstellungen erprobt (vgl. die schon oben erwähnten Beispiele aus den Abschnitten 5.1—5.3). Man sollte freilich wesentliche Nachteile der Methode nicht übersehen. Sie ist nicht konstruktiv und setzt die a-priori-Kenntnis der Periode voraus; Eindeutigkeits- und Stabilitätsfragen bleiben ausgeklammert. Der restliche Teil des Buches ist gewissermaßen der komplementären Situation gewidmet und unabhängig vom ersten Teil lesbar. Die Betrachtungen basieren ausschließlich auf Iterationsverfahren und betreffen nur den schwach-nichtlinearen Fall (kleine Parameter). Dafür sind sie auch auf fastperiodische Lösungen anwendbar, und es können die Existenzaussagen durch weitere Informationen (Eindeutigkeit, Stabilität) ergänzt werden. Aus diesem Grunde ist die in Kap. 6 und 7 behandelte Mittelwertmethode — trotz ihrer Gebundenheit an „kleine" Parameter — wohl nach wie vor das für die Anwendungen wichtigste analytische Verfahren der nichtlinearen Mechanik. Ihr Kernstück ist die Transformation eines Systemes auf die Normalform (6.2.30) (p. 132), die dann iterativ in aller Ausführlichkeit weiterbehandelt wird. Es sei angemerkt, daß es zu diesem Iterationsverfahren Alternativen gibt, die ein Mehr an Einsicht bringen würden. Was sich Ref. an dieser Stelle gewünscht hätte, wäre ein stärkeres Eingehen auf die eigentliche Problematik der Mittelwertmethode, nämlich die Charakterisierung derjenigen Systeme, die auf die Normalform so transformierbar sind, daß die Matrizen A, S keine rein-imaginären Eigenwerte besitzen. Daß dies auch in Fällen geht, wo zunächst die linearisierte Gleichung „kritisch" ist — wie im Abschnitt 6.4 gesagt wird — macht ja das eigentlich Interessante an diesem Verfahren aus.

Im Kapitel 8 wird die gleiche Aufgabe wie in den beiden vorangehenden Kapiteln unter anderen Voraussetzungen und mit der von J. MOSER für die Konstruktion von invarianten Kurven entwickelten Methode behandelt. Das Hauptresultat findet sich versteckt auf p. 213 und ist insofern etwas enttäuschend, als es nicht auf eine gegebene, sondern nur auf eine ihr zugeordnete modifizierte Gleichung anwendbar ist. Die Modifikation ist Nebenprodukt eines Iterationsprozesses und a-priori nicht bekannt. Die konkrete Substanz dieses Resultates, insbesondere im Hinblick auf Anwendungen, wird nicht näher diskutiert, was sicher ein Desideratum darstellt. So bleibt der Anreiz zur Lektüre des letzten Kapitels wohl auf die Tatsache beschränkt, daß hier eine gut lesbare Einführung in die von KOLMOGOROFF, ARNOLD und J. MOSER begründete Technik der beschleunigten Konvergenz geboten wird. Wie überhaupt festzustellen ist, daß das Schwergewicht der Darstellung auf einer präzisen und geschickt dargebotenen Beschreibung jeder einzelnen der verschiedenen Beweistechniken liegt, daß ein Vergleich der verschiedenen Zugänge aber nicht versucht wird. Eine umfassende vergleichende Würdigung der gebräuchlichen analytischen Verfahren in der nicht-linearen Mechanik steht auch nach dem Erscheinen des vorliegenden sonst sehr lesenwerten Büchleins noch aus.

Würzburg H. W. KNOBLOCH

van der Houwen, P. J., Construction of Integration Formulas for Initial Value Problems. Amsterdam-New York-Oxford. North-Holland. 1976. XII, 276 S., Dfl. 85.—. US $ 33.95 (Applied Mathematics and Mechanics).

Der Verf. behandelt in seiner Monographie die wichtigsten Problemstellungen der Konstruktion von Näherungsverfahren für die numerische Lösung von gewöhnlichen Anfangswertproblemen, wobei hauptsächlich auf steife Differentialgleichungssysteme eingegangen wird. Die Näherungsverfahren sind immer Differenzenverfahren. Besonderes Augenmerk wird in diesem Zusammenhang auf die zentralen Punkte in diesem Gebiet, Konsistenz, Konvergenz und besonders auf die Stabilität, gelegt. Die behandelten Verfahren gehören zu den TAYLOR-, RUNGE-KUTTA- und verallgemeinerten RUNGE-KUTTA-Verfahren. Einen breiten Raum nimmt die Konstruktion von Näherungsverfahren mit einer angepaßten Stabilitätsfunktion ein, wobei die Verfahren von LINIGER und WILLOUGHBY, Verfahren des Verfassers u. a. besprochen werden. Auf die Wahl der Schrittweite wird eingegangen, außerdem wird über numerische Experimente berichtet. Die einzelnen Kapitel behandeln: Einführung in die Problemstellung, Einzelschrittverfahren, Mehrschrittverfahren und Stabilitätspolynome.

Die Monographie ist gut lesbar, für ihr Studium sind nur wenige mathematische Vorkenntnisse erforderlich. Das wohlgelungene Buch kann allen sehr empfohlen werden, die gewöhnlichen Anfangswertprobleme mit einer Rechenmaschine numerisch lösen wollen.

Berlin F. GRUND

Bosch, K., Elementare Einführung in die Wahrscheinlichkeitsrechnung, Braunschweig. Vieweg. 1976. X, 192 S.

Vorliegendes Taschenbuch gehört in die Reihe „rororo vieweg", deren Programm die Gebiete Mathematik, Physik, Chemie, Biologie umfaßt und die durch Bände „Basiswissen" abgerundet wird, in denen fachübergreifende Themen und wissenschaftstheoretische Grundlagen behandelt werden. Die „Elementare Einführung in die Wahrscheinlichkeitsrechnung" von KARL BOSCH ist — und das ist schon bemerkenswert — als ein Band des Themenplanes „Basiswissen" ausgewiesen. Dabei handelt es sich um eine elementare Einführung in die Wahrscheinlichkeitsrechnung z.B. in dem Sinne, daß auf den Isomorphiesatz von M. H. STONE nicht eingegangen wird, daß auf einen maßtheoretischen Aufbau verzichtet wird und daß diskrete und stetige Zufallsgrößen getrennt behandelt werden. Diese Einführung ist aber recht umfassend; es wird ein ziemlich komplettes Standardprogramm zur Wahrscheinlichkeitsrechnung bis hin zu den Gesetzen der großen Zahlen und dem zentralen Grenzwertsatz — letzteren ohne Beweis — geboten. Am Ende gibt es noch einen kleinen Ausblick auf die Probleme der mathematischen Statistik, wie sie im Band „Angewandte mathematische Statistik" derselben Reihe und desselben Autors behandelt werden.

Der Anhang enthält neben den Lösungen der Übungsaufgaben eine Tafel der Verteilungsfunktion der standardisierten Normalverteilung und ein — allerdings nicht sonderlich umfassendes — Verzeichnis über weiterführende Literatur.

M. E. wird der Autor den Ansprüchen, wie sie sich aus den allgemeinen Zielen der Taschenbuchreihe „rororo vieweg" ergeben, in hohem Maße gerecht. Ohne vom Leser mehr mathematische Kenntnisse zu verlangen, als man bei Hochschulreife erwarten darf, wird die Wahrscheinlichkeitsrechnung in mindestens dem Ausmaß aufgebaut, das für eine solide mathematische Formulierung der Probleme der mathematischen Statistik, ihrer Verfahren und deren Begründung erforderlich ist. Besonders möchte ich schließlich noch auf Anzahl und Vielfalt der Beispiele und Aufgaben hinweisen. Dabei dienen die Beispiele häufig auch zur Vorbereitung, Motivation und zum Plausibelmachen von Definitionen und Aussagen — und dies erleichtert dem Anfänger in der Wahrscheinlichkeitsrechnung die Sache doch bekanntlich sehr.

Dresden G. Maibaum

Bosch, K., Angewandte mathematische Statistik, Braunschweig. Vieweg & Sohn. 1976. 210 S., 41 Abb.

Mit vorliegendem Buch wird die Taschenbuchreihe „rororo vieweg", die vom Rowohlt-Taschenbuch-Verlag in Zusammenarbeit mit dem Verlag Vieweg herausgegeben wird, durch einen weiteren Band ergänzt. Dabei ist die „Angewandte mathematische Statistik" von Karl Bosch — wie auch die „Elementare Einführung in die Wahrscheinlichkeitsrechnung" desselben Autors — als Band des Themenplanes „Basiswissen" eingeordnet. Das erklärte Ziel des Verfassers ist es, die Verfahren und Methoden der mathematischen Statistik nicht nur darzustellen und deren konkrete Anwendung an Beispielen zu zeigen, sondern diese Verfahren und Methoden immer auch zu begründen, die Vorgehensweise zu motivieren, die Begriffsbildungen plausibel zu machen und damit schließlich auch Verständnis zu erreichen für die Möglichkeiten und Grenzen der mathematischen Statistik. Es ist dem Autor zu bescheinigen, daß dieses Anliegen in hohem Maße erfüllt wird und daß das kleine Buch dem angesprochenen Leserkreis — das sind in der Hauptsache wohl Studenten, die während des Studiums irgendwie mit dem Fach „Statistik" in Berührung kommen — sehr nützlich sein kann. Es bietet ein ausgewogenes Standardprogramm der mathematischen Statistik, wobei in Tabellen — teils im Text, teils am Ende des Buches — ausreichend viele Quantile in Frage kommender Verteilungen angegeben werden, um bei der praktischen Anwendung der dargestellten Verfahren auch zu einem konkreten Ergebnis oder einem zahlenmäßigen Resultat zu gelangen. Für die Lektüre sind — was allgemein-mathematische Kenntnisse anbelangt — solche Kenntnisse ausreichend, wie sie bei Hochschulreife erwartet werden dürfen; eine Vorbildung auf dem Gebiet der Wahrscheinlichkeitsrechnung ist in dem Umfange des schon genannten rororo vieweg-Taschenbuches „Elementare Einführung in die Wahrscheinlichkeitsrechnung" für das Verständnis recht nützlich und bis zu einem hohen Grade auch notwendig.

Es ist klar, daß in dem kleinen Buch (knapp 190 Seiten Text) nicht alle Aussagen bewiesen werden. Die Darstellung ist bemerkenswert klar, übersichtlich gegliedert, durch Beispiele gut aufgelockert. Zu den Tests gibt es neben der praktischen Entscheidungsregel häufig Hinweise, die sich aus der Erfahrung ergeben haben und denen bei der Anwendung Beachtung geschenkt werden sollte.

Zusammenfassend möchte ich feststellen, daß die „Angewandte mathematische Statistik" von Karl Bosch unter den vergleichbaren deutschsprachigen Lehrbüchern — das Literaturverzeichnis gibt hierüber allerdings keinen umfassenden Überblick — m. E. einen sehr guten Platz einnehmen wird.

Dresden G. Maibaum

Ehrhart, E., Polynômes arithmétiques et Méthode des Polyèdres en Combinatoire. Basel-Stuttgart. Birkhäuser Verlag. 1977. 165 S., sFr. 32.– (ISNM 35).

Der Autor ist bekannt als Fachmann für diophantische Gleichungen, Zahlengeometrie und damit zusammenhängende kombinatorische Fragestellungen. Die Fülle seiner Veröffentlichungen dazu, insbesondere seine im J. reine angew. Math. **226** 1–29 (1967), **227**, 25–49 (1967) publizierte Thèse "Sur un problème de géométrie diophantienne linéaire" und eine weitere Arbeit "Sur le nombre de solutions des systèmes diophantiens linéaires" (U.E.R. Math. Strasbourg 1972) sowie Beiträge von anderen Autoren und eigene neue Resultate sind im vorliegenden Band in eine systematische Darstellung und Diskussion des Forschungsgegenstandes des Autors eingebettet. Der Wert der dabei bereitgestellten, vom Verfasser entwickelten Hilfsmittel wird bei Anwendungen in verschiedenen Problemkreisen überzeugend demonstriert. Damit werden Richtungen für erfolgversprechende weitere Untersuchungen gewiesen. Hervorgehoben seien der nützliche Begriff des arithmetischen Polynoms: Polynom in n, dessen Koeffizienten in n periodische Zahlen sind — für die der Autor eine zweckmäßige Schreibweise einführt — und das für ganzzahlige n ganzzahlige Werte annimmt, weiter ein bemerkenswertes Reziprozitätsgesetz für die Zahl der ganzzahligen Gitterpunkte in einem k-dimensionalen Polyeder und die im Titel genannte abzählende Polyedermethode, die an Magischen Quadraten, Partitionen, Verteilungsproblemen u. a. demonstriert wird. Überhaupt machen durchgerechnete Beispiele einen großen Teil des Textes aus. Sie lockern diesen vorteilhaft auf, indem sie auf die Einführung neuer Hilfsmittel vorbereiten oder deren Handhabung veranschaulichen. Der Verfasser setzt beim Leser ein mathematisches Allgemeinwissen, aber keine Spezialkenntnisse voraus. — Inhalt: 1re partie: Les polynômes arithmétiques. Ch. I. Nombres périodiques, p. 7. Ch. II. Polynômes arithmétiques, p. 12. 2me partie: Polyèdres et réseaux. Ch. III. Polyèdres homothétiques, p. 47. Ch. IV. Polygones paramétriques, p. 61. 3me partie: La méthode des polyèdres en Combinatoire. Ch. V. Problèmes, p. 74. Ch. VI. Systèmes diophantiens linéaires, p. 99. Appendice, pp. 141–158.

Berlin W. Wessel

Sampson, J. R., Adaptive Information Processing. An Introductory Survey. Berlin-Heidelberg-New York. Springer-Verlag. 1976. X, 214 S., 83 Abb., DM 36,50. US $ 14.80 (Reihe: Texts and Monographs in Computer Science).

Das Buch faßt einen sehr weiten Bogen der Informationsprozesse. Es beginnt mit Grundlagen der Informations- und Codierungstheorie sowie den endlichen, Turing- und Zellularautomaten. Bei der biologischen Information werden die wesentlichen genetischen (in der biologischen Zelle ablaufenden) und neuronalen (in Gehirnen ablaufenden) Prozesse beschrieben, sowie auf ihre Rechnersimulation eingegangen. Im dritten Teil werden wichtige Gebiete der künstlichen Intelligenz, wie Zeichenerkennung, Spieltheorie, Theorembeweisen, Problemlösen und natürliche Sprachen behandelt. Dieses Spektrum läßt zwar einen guten Allgemeinüberblick zu, kann aber auf keinem Gebiet in die Tiefe führen. Dies ist wohl auch nicht vom Autor und Verlag angestrebt. Geplant ist das Buch nämlich für nichtgraduierte Rechentechniker. Es kann jedoch wohl eher als ein gut ausgewähltes und verständliches Werk der Informationsprozesse auf populärwissenschaftlichem Gebiet (im besten Sinne des Wortes) gelten und dürfte daher für einen breiten Leserkreis von Interesse sein.

Berlin H. Völz

In der DDR beziehbar im Rahmen des Kontingents.

EINGEGANGENE BÜCHER

Bei der Schriftleitung sind folgende Bücher eingegangen (ausführliche Besprechung bleibt vorbehalten).

Trachtenbrot, B. A., Algorithmen und Rechenautomaten. Berlin. Deutscher Verlag der Wissenschaften. 1977. 207 S., 55 Abb., 3 Tab. DM 17,80. Übersetzung aus dem Russischen.

Federn, K., Auswuchttechnik. Bd. 1: Allgemeine Grundlagen — Meßverfahren und Richtlinien. Berlin-Heidelberg-New York. Springer-Verlag. 1977. Geb. DM 88,—; US $ 38.80. 195 Abb. XVIII, 449 Seiten.

Gähler, W., Grundstrukturen der Analysis I. Berlin. Akademie-Verlag. 1977. VIII, 412 Seiten. 48,— M.

Dem'janov, W. F. / Malozemov, W. N., Einführung in Minimax-Probleme. Leipzig. Geest & Portig K.-G. 1975. 288 Seiten, 37 Abb., Leinen 49,50 M.

Ameling, W., Planung und Entwicklung einer programmierbaren digitalen Integrieranlage. Forschungsberichte des Landes Nordrhein-Westfalen. 1976. Opladen. Westdeutscher Verlag GmbH. 177 Seiten. DM 32,—.

Müller, P. C. / Schiehlen, W. O., Lineare Schwingungen. Wiesbaden. Akademische Verlagsgesellschaft. 1976. X, 363 Seiten.

Forsythe, G. E. / Malcolm, M. A. / Moler, C. B., Computer Methods for Mathematical Computations. Englewood Cliffs, New Jersey 07632. Prentice Hall, Inc., 1977. XI, 259 S.

Kupka, I. / Wilsing, N., Dialogsprachen. Stuttgart. B. G. Teubner. 1975. 168 S., 10 Abb., 3 Tab., DM 19,80. Teubner Studienbücher, LAMM Bd. 32.

Wirth, N., Algorithmen und Datenstrukturen. Stuttgart. B. G. Teubner. 1975. 376 S., 93 Fig., 30 Tab. Kart. DM 26,80. Teubner Studienbücher, LAMM Bd. 31.

Gruber, C. / Hintermann, A. / Merlini, D., Group Analysis of Classical Lattice Systems. Berlin-Heidelberg-New York. Spinger-Verlag. 1977. XIV, 326 S., 49 Abb., DM 31,—. US $ 13.70 (Lecture Notes in Physics 60).

Lamperti, J., Stochastic Processes. A Survey of the Mathematical Theory. Berlin-Heidelberg-New York. Springer-Verlag 1977. XVII, 266 S., DM 21,40. US $ 9.80 (Applied Mathematical Sciences 23).

Chadan, K. / Sabatier, P. C., Inverse Problems in Quantum Scattering Theory. Berlin-Heidelberg-New York. Springer-Verlag. 1977. XII, 344 S., 24 Abb., DM 67,70. US $ 29.80 (Texts and Monographs in Physics).

Provis, D. G. / Radok, R., Waves on Water of Variable Depth. Berlin-Heidelberg-New York. Springer-Verlag. 1977. Canberra. Australian Academy of Science. 231 S., 134 Abb., DM 24,80. US $ 11.00.

Eiselt, H. A. / von Frajer, H., Operations Research Handbook. Standard Algorithms and Methods. Berlin-New York. de Gruyter. 1977. 398 S., DM 74,—.

Smith, J. M., Mathematical Modelling & Digital Simulation for Engineers & Scientists. New York-London-Sidney-Toronto. John Wiley & Sons. 1977. 332 S., £ 14.50. $ 26.—.

Nakamoto, K., Infrared Spectra of Inorganic and Coordination Compounds, Second Edition. New York-London-Sydney-Toronto. J. Wiley & Sons. 1977. 338 S., £ 18.75. $ 32.00.

Marti, J. T., Konvexe Analysis, Basel-Stuttgart. Birkhäuser Verlag. 1977. XI. 273 S., SFr 48,—. (Mathematische Reihe 54).

Salençon, J., Applications of the Theory of Plasticity in Soil Mechanics, Chichester-New York-Brisbane-Toronto. J. Wiley & Sons. 1977. XII, 158 S., £ 10.00. $ 19.00.

Wunsch, G., Zellulare Systeme, Berlin. Akademie-Verlag. 1977. 150 S., M 8,—. (WTB Mathematik—Physik 194).

Kaiser, H., Numerische Mathematik und Rechentechnik I, Berlin. VEB Deutscher Verlag der Wissenschaften. 1977. 216 S., 60 Abb., M 16,80. (Mathematik f. Lehrer, Bd. 9).

Beer, K., Lösung großer linearer Optimierungsaufgaben, Berlin. VEB Deutscher Verlag der Wissenschaften. 1977. 254 S., 15 Abb., M 40,—.

Schneider, W., Mathematische Methoden der Strömungsmechanik, Braunschweig. Vieweg. 1978. IX, 261 S., 94 Abb., DM 39,80.

Koch, G. / Rembold, U., Einführung in die Informatik, für Ingenieure und Naturwissenschaftler, Teil 1, München-Wien. Carl Hanser Verlag. 1977. 224 S., 184 Abb., DM 28,—. (Studienbücher d. Techn. Wissenschaften)

Radaj, D. / Schilberth, G., Kerbspannungen an Ausschnitten und Einschlüssen, Düsseldorf. Deutscher Verlag für Schweißtechnik. 1977. 217 S., 152 Abb., DM 98,—. (Fachbuchreihe Schweißtechnik 69).

Oswatitsch, K., Spezialgebiete der Gasdynamik, Schallnähe, Hyperschall, Tragflächen, Wellenausbreitung, Wien-New York. Springer-Verlag. 1977. IX, 377 S., 161 Abb., S 1325,—. DM 192,—.

Villaggio, P., Qualitative Methods in Elasticity, Leyden. Noordhoff International Publishing. 1977. IX, 692 S., Dfl. 270.00. $ 112.50.

Temam, R., Navier-Stokes Equations, Amsterdam-New York. North-Holland Publishing Company. 1977. XVI, 454 S., US $ 45.00. Dfl. 125.00. (Studies in Mathematics and its Applications 2)

Bochmann, D. / Roginskij, V. N., Dynamische Prozesse in Automaten, Berlin. VEB Verlag Technik. 1977. 219 S., M 30,—.

Schlageter, G. / Stucky, W., Datenbanksysteme: Konzepte und Modelle, Stuttgart. B. G. Teubner, 1977. 259 S., DM 22,80. (LAMM 37).

Richter, L., Betriebssysteme, Stuttgart. B. G. Teubner. 1977. 152 S., DM 22,80. (LAMM 29).

Kohlas, J., Stochastische Methoden des Operation Research, Stuttgart. B. G. Teubner. 1977. 192 S., DM 24,80. (LAMM 40).

Todd, J., Basic Numerical Mathematics. Vol. 2: Numerical Algebra. Basel-Stuttgart. Birkhäuser Verlag. 1977. 216 S., (ISNM 22).

Werum, W. / Windauer, H., Pearl, Process and Experiment Automation Realtime Language. Braunschweig. Vieweg. 1978. 195 S., DM 29,80.

Stange, K., Bayes-Verfahren, hrsgeg. von Deutler, T. / Wilrich, P.-T., Berlin-Heidelberg-New York. Springer-Verlag. 1977. VIII, 312 S., 36 Abb., DM 39,—. US $ 17.20 (Reihe: Hochschultext).

Gaede, K.-W., Zuverlässigkeit, Mathematische Modelle. München-Wien. Carl Hanser Verlag. 1977. 254 S., 68 Abb., DM 48,—.

Kramer, H., Assembler IV, Supplement zum Lernprogramm. Berlin-Heidelberg-New York. Springer-Verlag. 1977. XI, 146 S., 207 Abb., DM 21,80. US $ 10.10 (Heidelberger Taschenbücher 189).

Schmidt, B., GPSS-FORTRAN. Einführung in die Simulation diskreter Systeme mit Hilfe eines FORTRAN-Programmpaketes. Berlin-Heidelberg-New York. Springer-Verlag. 1977. IX, 298 S., 27 Abb., DM 38,—. US $ 16.80 (Informatik-Fachberichte 6).

Nagel, H.-H. (Hrsg.), Digitale Bildverarbeitung / Digital Image Processing, GI/NTG Fachtagung — München, 28.—30. 3. 1977. Berlin-Heidelberg-New York. Springer-Verlag. 1977. XI, 328 S., 180 Abb., DM 31,—. US $ 13.70 (Informatik-Fachberichte 8).

Freund, H. / Sorger, P., Aussagenlogik und Beweisverfahren. Stuttgart. B. G. Teubner. 1974. 136 S., 108 Abb., DM 16,80 (Reihe: Mathematik für Lehrausbildung).

Müller, P. C., Stabilität und Matrizen. Matrizenverfahren in der Stabilitätstheorie linearer dynamischer Systeme. Berlin-Heidelberg-New York. Springer-Verlag. 1977. VII, 220 S., 18 Abb., DM 38,—. US $ 16.80.

Szabó, J. / Roller, B., Anwendung der Matrizenrechnung auf Stabwerke. Budapest. Akadémiai Kiadó. 1978. 347 S., Ft. 418,—.

Gander, W. / Molinari, L. / Svecová, H. (Hrsg.), Numerische Prozeduren aus Nachlaß und Lehre. Von Prof. Heinz Rutishauser, Basel-Stuttgart. Birkhäuser Verlag. 1977. 127 S., (ISNM 33).

Jansen, J. K. M., Simple-Periodic and Non-Periodic Lamé Functions. Amsterdam. Mathematisch Centrum. 1977. 104 S., Dfl. 13,— (MC Tracts 72).

Moffatt, H. K., Magnetic Field Generation in Electrically Conducting Fluids. Cambridge-London-New York-Melbourne. Cambridge University Press. 1978. 343 S., £ 15.50 A H/B.

Koiter, W. T., Theoretical and Applied Mechanics. Amsterdam-New York-Oxford. North Holland Publishing Company. 1977. 491 S., Dfl. 120.—. US $ 48.95.

NACHRICHTEN

Im Rahmen der Wissenschaftlichen Tage 1979 der TH Otto von Guericke Magdeburg finden statt:

V. Symposium Verformung und Bruch vom 4. bis 6. 9. 1979

Das Symposium soll zur Entwicklung der interdisziplinären Zusammenarbeit von Festkörperphysikern, Werkstoffwissenschaftlern und Festkörpermechanikern bei der Lösung von Aufgaben der Grundlagen- und angewandten Forschung durch die Behandlung folgender Themenkomplexe beitragen:

- Physikalische Grundlagen der Verformung und des Bruchs
- Weiterentwicklung und Anwendung der experimentellen Bruchmechanik
- Methoden der Festkörpermechanik zur Analyse von Verformungs- und Bruchvorgängen.

2. Tagung Strömungsmechanik vom 3. bis 6. 9. 1979

Gemeinsam mit der Mathematischen Gesellschaft der DDR und der Akademie der Wissenschaften der DDR, Zentralinstitut für Mathematik und Mechanik, veranstaltet, soll die Tagung der Entwicklung der interdisziplinären Zusammenarbeit von Mathematikern, Wissenschaftlern der theoretischen und angewandten Mechanik sowie Vertretern der Praxis bei der Lösung aktueller strömungstechnischer Probleme aus den Themenkomplexen

- Mehrphasen- und Mehrkomponentenströmungen
- Turbulente und rheologische Strömungen
- Angewandte strömungstechnische Probleme des Maschinenbaus

dienen.

Interessenten können nähere Informationen über das Tagungsbüro der TH Otto von Guericke, 301 Magdeburg, Postfach 124 (Tel. 592209) anfordern.

The **Second International Symposium on Innovative Numerical Analysis in Applied Engineering Science** will be held June 16—20, 1980 at the Ecole Polytechnique de Montreal, Canada. Papers are sollicited in all areas of engineering science such as solid and fluid mechanics, electrical engineering, acoustics, etc... with emphasis on "non-standard" numerical analysis. Papers on innovative numerical analysis which emphasize technology transfer topics, hybridization of analysis methods and "new" technology are strongly encouraged.

Extended abstracts of approximately 1000 words, in French or English, are required by March 15, 1979. Notification of acceptance will be around June 15, 1979. Final accepted papers of up to 10 pages will be required by December 15, 1979 for publication in the symposium proceedings. Five copies of each abstract should be sent to Dr. A. Chaudouet, CETIM, Boite Postale No. 67, 60304 Senlis Cedex, France (for Europa-Africa-Middle East) or to Dr. T. Cruse, PRATT AND WHITNEY AIRCRAFT, East Hartford, Conn. 06108, U.S.A. (for America-Japan-Australia-New Zealand).